Engineering Materials

For further volumes:
http://www.springer.com/series/4288

Michael D. Kotsovos

Compressive Force-Path Method

Unified Ultimate Limit-State Design
of Concrete Structures

 Springer

Michael D. Kotsovos
School of Civil Engineering
  Reinforced Concrete Structures
National Technical University of Athens
Athens
Greece

ISSN 1612-1317   ISSN 1868-1212   (electronic)
ISBN 978-3-319-34542-0   ISBN 978-3-319-00488-4   (eBook)
DOI 10.1007/978-3-319-00488-4
Springer Cham Heidelberg New York Dordrecht London

Printed on acid-free paper

Springer is part of Springer Science+Business Media (www.springer.com)

In the memory of Milija N. Pavlovic

Preface

Current Design Philosophy

Current codes of practice for the design of structures have been developed within the context of the limit-state philosophy: A structure or member is first designed so as to exhibit specified performance after attaining its load-carrying capacity, i.e. when its *ultimate limit state* is reached; the design is complemented or even revised during a process of checking whether the structure or member exhibits the desired behavioural characteristics under service conditions, i.e. at the *serviceability limit state*.

Anticipated Benefits

The adoption of the limit-state philosophy as the basis of current codes of practice for the design of concrete structures expresses the conviction that this philosophy is capable of leading to safer and more economical design solutions. After all, designing a structural concrete member to its ultimate limit state requires the assessment of the load-carrying capacity of the member and this provides a clearer indication of the margin of safety against collapse. At the same time, the high internal stresses which develop at the ultimate limit state result in a reduction of the member cross-section and the amount of reinforcement required to sustain internal actions. (Admittedly, the latter economy and, of course, safety itself are dependent on the actual safety factor adopted; nevertheless, the more accurate estimate of the true failure load provides an opportunity to reduce the uncertainties reflected in the factor of safety in comparison with, say, elastic design solutions).

Shortcomings

In contrast to the above expectations for more efficient design solutions, there have been instances of concrete structures which have been reported to have suffered unexpected types of damage under earthquake action, whereas a number of attempts to investigate experimentally whether or not the aims of limit-state philosophy for safety and economy are indeed achieved by current codes of practice have yielded conflicting results. Experimental evidence has been published that describes the behaviour of a wide range of structural concrete members (such as, for example, beams, columns, beam-column joints, walls, etc.) for which current methods for assessing structural performance yield predictions exhibiting excessive deviations from the true behaviour as established by experiment. In fact, in certain cases the predictions underestimate considerably the capabilities of a structure or member—indicating that there is still a long way to go in order to improve the economy of current design methods—while in other cases the predictions are clearly unsafe as they overestimate the ability of a structure or member to perform in a prescribed manner, in spite of the often excessive amount of reinforcement specified; and this provides an even more potent pointer to the fact that the rational and unified design methodology is still lacking in structural concrete. The lack of such a methodology is also reflected in the complexity and the segmented per structural element and performance requirement nature of the code specifications.

Need for Revision of Design Methods

As it will become apparent in the following chapter, the investigation of the causes of the above shortcomings led to the conclusion that the conflicting predictions are due to the inadequacy of the theoretical basis of the design methods which are used to implement the limit-state philosophy in practical design, rather than the unrealistic nature of the aims of the design philosophy as such. In fact, it was repeatedly shown that the fundamental assumptions of the design methods which describe the behaviour of concrete at both the material and structure levels were adopted as a result of misinterpretation of the available experimental information and/or use of concepts which, while working well for other materials (e.g. steel) or regimes (e.g. elastic behaviour), are not necessarily always suitable to concrete structures under ultimate-load conditions, i.e. at the ultimate limit state. Therefore, it becomes clear that the theoretical basis of current design methods requires an extensive revision if the methods are to consistently yield realistic predictions as a result of a rational and unified approach.

Proposed Revisions

Such a revision has been the subject of comprehensive research work carried out by Kotsovos and Pavlovic over the past three decades. This was done concurrently at two levels. One of these—the 'higher' level—was based on formal finite-element (FE) modelling of structural concrete with realistic material properties and behaviour as its cornerstone: a large part of the ensuing results are contained in the book by Kotsovos and Pavlovic, *Structural Concrete*, published in 1995; those obtained after 1995 are intended to be included in a new edition of the book. At the second—the 'lower', level an attempt was made to reproduce the essential results of complex numerical computations by means of much simpler calculations which would require no more effort than is the case with current code provisions. The latter approach was deemed necessary because, although the Kotsovos and Pavlovic's FE model has proved useful as a consultancy tool for the design, redesign, assessment and even upgrading of reinforced concrete structures, the fact remains that most design offices still rely on simplified calculation methods which, if not quite 'back-of-the-envelope' stuff, are quick, practically hand-based (or easily programmable), provide (or claim to provide) a physical feel for the problem, and, of course, conform to the simple methodology of code regulations.

Compressive-Force Path Method

The alternative methodology at this level, which stems from the Kotsovos and Pavlovic's work and is the subject of their book *Ultimate Limit-State Design of Concrete Structures*, published in 1999, and which provided the basis for a new, improved design approach for the implementation of the limit-state philosophy into the practical design of concrete structures, involves, on the one hand, the identification of the regions of a structural member or structure at its ultimate limit state through which the external load is transmitted from its point of application to the supports, and, on the other hand, the strengthening of these regions so as to impart to the member or structure desired values of load-carrying capacity and ductility. As most of the above regions enclose the trajectories of internal compressive actions, the new methodology has been termed the 'compressive-force path' (CFP) method. In contrast to the methods implemented in current codes of practice, the proposed methodology is fully compatible with the behaviour of concrete (as described by valid experimental information) at both the material and structure levels capable of producing design solutions that have been found to satisfy the code performance requirements in all cases investigated.

It may also be of interest to note that, although the CFP method might appear, at first sight, to be a rather unorthodox way of designing structural concrete, it is easy, with hindsight, to see that it conforms largely to the classical design of masonry structures by Greek and Roman Engineers. These tended to rely greatly on arch action—later expressed (and extended) through the Byzantine dome and the Gothic vaulting. Now, such a mechanism of load transfer may seem largely irrelevant for a beam exhibiting an elastic response. However, for a cracked reinforced concrete girder close to failure the parallel with an arch-and-tie system reveals striking similarities between the time-honoured concept of a compressive arch and the newly proposed CFP method.

Revised Compressive-Force Path Method

Since the publication of the book by Kotsovos and Pavlovic in 1999, the first of the authors has focused his efforts into generalising the CFP method so as to extend its application into the whole range of practical cases covered by current codes of practice for the design of RC structures, earthquake-resistant RC structures inclusive. This was achieved by replacing the failure criteria with simple expressions which were derived from first principles without the need for calibration through the use of experimental data on structural concrete behaviour. The implementation of these criteria into the CFP method not only simplified the assessment of the strength characteristics of structural concrete, but also led to a drastic revision of the method and extended its use into the whole range of structural elements of common RC structures.

Present Book

The aim of the present book, therefore, is to introduce to designers the revised version of the 'Compressive-Force Path' Method. Such an introduction not only includes the description of its underlying theoretical concepts and their application in practice but, also, presents the causes which led to the need for a new design methodology for the implementation of the limit-state philosophy into practical structural design together with evidence—both experimental and analytical—supporting its validity.

The book is divided into eight chapters. Chapter 1 presents in a unified form and discusses all available information on the conflict between the concepts underlying current code provisions and the causes of the observed and/or measured structural behaviour. The information presented in Chap. 1 is summarised in Chap. 2 so as to form the theoretical basis of the proposed design methodology. The latter forms the subject of Chaps. 3–7 which concentrate not only on its description but, also, on its implementation into practical design and the presentation of evidence of its validity.

More specifically, Chap. 3 presents the physical model which underlies the application of the methodology for the design of simply supported beams, together with failure criteria capable of providing a realistic prediction of the load-carrying capacity for all types of behaviour characterising such beams. The physical model and the failure criteria presented in Chap. 3 are used for the development of the design method discussed in Chap. 4; this method is extended so as to apply for punching, as described in Chap. 5, and for any structural concrete configuration comprising beam, column or wall elements, as described in Chap. 6. Chapter 7 demonstrates that the proposed methodology is also applicable to the design earthquake-resistant RC structures without the need of the modifications normally required by the methods adopted by current codes for the design of RC structures under normal loading conditions. Finally, the presentation of typical examples of the application of the proposed method in design forms the subject of Chap. 8.

The author is fully aware, of course, that code tenets cannot be ignored by the majority of designers, not only because of legal implications but, more positively, many guidelines accumulate vast practical experience regarding the detailing of a wide range of reinforced concrete structures. Nevertheless, there are clearly problems for which code guidelines are less successful, and such difficulties need to be addressed. The present book is intended to address such problems. Ultimately, however, it is up to the experienced engineer, as well as the young graduate or student well acquainted with present-day code rules, to decide whether or not ideas contained in this book do, in fact, provide a rational alternative to the design of structural concrete members.

Acknowledgments

The author wishes to express his gratitude to Dr. Jan Bobrowski (of Jan Bobrowski and partners, and Visiting Professor at Imperial College London) for the many discussions he had with him on various aspects of structural concrete design: his research work has been strongly influenced by Jan's design philosophy and achievements, and this is certainly reflected in the method proposed in this book.

The author is also indebted to past students who contributed to the shaping of the present unifying method for the design of concrete structures. Particular mention must be made of Dr. Gerasimos Kotsovos who contributed to the development and verification of the failure criteria which formed the basis for the revision of the CFP method. Dr. Gregoria Kotsovou and Dr. Charis Mouzakis extended the application of the method to the design of earthquake-resistant beam-column joints, whereas Dr. Emmanuel Vougioukas and Dr. Demetrios Cotsovos verified the validity of the method for the case of two-span linear elements and structural walls, respectively.

Michael D. Kotsovos

Contents

Chapter 1
Reappraisal of Concepts Underlying Reinforced-Concrete Design

1.1 Introduction

Since the mid-eighties, there has been an increasing amount of experimental evidence which shows that many of the concepts underlying current-code provisions for the design of reinforced-concrete (RC) structures are in conflict with fundamental properties of concrete at both the material and the structure levels [1]. More recently, it has been shown that this conflict has been the cause of the unexpected (in accordance with current codes such as, for example, ACI318 [2] and EC2 [3]/EC8 [4]) damage suffered at mid height by the vertical (column and structural-wall) elements of RC buildings during the 1999 Athens earthquake [5]. In fact, this finding has been confirmed from the results of tests that reproduced this type of damage under controlled laboratory conditions [6–8]. Moreover, the latter tests not only revealed additional weaknesses of the provisions of current codes for earthquake-resistant design [9], but also indicated that it is, in fact, possible to obtain design solutions that satisfy the performance requirements of the codes through the use of alternative design approaches that allow for a realistic description of structural-concrete behaviour [6–9].

To this end, the aim of the present chapter is to collate all available information on the conflict between the concepts underlying current-code provisions and the causes of the observed and/or measured structural behaviour, and present it in a unified form. Such information involves fundamental aspects of RC design which are associated with, not only flexural and shear design, but also with elements of the design of earthquake-resistant RC structures such as, for example, the design of hoop reinforcement for the "critical regions" [2–4] and the regions of points of contraflexure (points, other than simple supports, along the span of a linear structural element, also known as points of inflection, where the bending moment is zero) [1, 5, 7]. Moreover, through the use of the above information, it will be demonstrated in subsequent chapters that the substitution of the concepts underlying the design methods adopted by current codes with alternative ones capable of providing a realistic description of structural-concrete behaviour, not only may

M. D. Kotsovos, *Compressive Force-Path Method*, Engineering Materials,
DOI: 10.1007/978-3-319-00488-4_1, © Springer International Publishing Switzerland 2014

improve the theoretical basis of RC design, but also simplify the design process and result in more efficient solutions without compromising the aims of structural design for safety, serviceability and economy.

1.2 Truss Analogy

1.2.1 Background

The truss analogy (TA), since its inception at the turn of the 20th century [10, 11], has always formed the basis of RC design. It became attractive for its simplicity and was first implemented in RC design through the permissible-stress philosophy. With the introduction of the limit-state philosophy in the 1970s, its use was extended for the description of the physical state of RC structures at their ultimate limit state by incorporating concepts such as strain softening [12], aggregate interlock [13, 14], dowel action [13, 15], etc. TA has remained to date the backbone of RC design, with more refined versions of it (in the form of the compression-field theory [16] and strut-and-tie models [17]) becoming increasingly popular.

The simplest form of such a truss describing the function of a beam-like element at its ultimate-limit state is shown in Fig. 1.1. In fact, this beam-like element is considered to start behaving as a truss once inclined cracking occurs, with the compressive zone and the flexural reinforcement forming the longitudinal struts and ties, respectively, the stirrups forming the transverse ties, whereas the cracked concrete of the element web is assumed to allow the formation of inclined struts.

1.2.2 Auxiliary Mechanisms of Shear Resistance

A characteristic feature of the implementation of the above model in current design practice is that the truss is often considered to sustain only a portion of the shear forces acting on a beam-like element; the remainder is considered to be sustained by the combined resistance to shear deformation offered by (a) the "uncracked" concrete of the compressive zone, (b) the "cracked" concrete of the tensile zone and (c) the flexural reinforcement, with the latter two auxiliary mechanisms of shear resistance being widely referred to as "aggregate interlock" [13] and "dowel action" [15], respectively.

Fig. 1.1 Truss modelling the function of an RC beam at its ultimate limit state

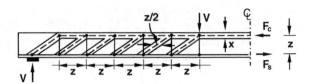

(a) Incompatibility with cracking mechanism

The mechanisms of "aggregate interlock" and "dowel action" can only be mobilized through the shearing movement of a crack's faces; and yet, such a movement is incompatible with the well-established cracking mechanism of concrete: a crack extends in the direction of the maximum principal compressive stress and opens orthogonally to its plane [18–20]. Therefore, a shearing movement cannot occur, and, as a result, the mechanisms of "aggregate interlock" and "dowel action" cannot be mobilised so as to contribute to the structural element's shear resistance.

(b) Incompatibility with mechanic's principles

Had it been possible for a shear movement to mobilize "aggregate interlock", its contribution to shear resistance could only be possible once the shear resistance of "uncracked" concrete in the compressive zone had been overcome. This is because, unlike "uncracked" concrete which exhibits strain-hardening behaviour, the behaviour of cracked concrete (within which aggregate interlock could only develop) is described by post-peak stress-strain material characteristics [20], and, hence, its stiffness is negligible, if any (as it will be discussed later), when compared with the stiffness of uncracked concrete. With such large difference in stiffness the contribution of "aggregate interlock" to the combined shear resistance can only be negligible.

(c) Incompatibility with observed structural behaviour

The validity of the concept of auxiliary mechanisms of shear resistance has been investigated experimentally by testing simply-supported beams under two-point loading (see Figs. 1.2 and 1.3) [21–23]. Figure 1.2a and b depict the geometric characteristics, together with the reinforcement details, of two types of beams with values of the shear span-to-depth ratio equal to approximately 1.5 and 3.3, respectively. The beams have the same geometric characteristics and longitudinal reinforcement but, with regard to the transverse reinforcement, they have beam classified as beams A, B, C, and D in Fig. 1.2a and beams A1, B1, C1 and D1 in Fig. 1.2b. The beams in Fig. 1.3 are similar to beams A in Fig. 1.2a or A1 in Fig. 1.2b, but they differ in the longitudinal reinforcement arrangement as indicated in the figure.

In accordance with current code provisions, for all beams, the load-carrying capacity corresponding to flexural capacity is significantly larger than the load-carrying capacity corresponding only to the contribution of the auxiliary mechanisms to shear resistance. Moreover, it should be noted that, for the beams in Fig. 1.2, the transverse reinforcement provided is sufficient, in accordance with current code provisions, to safeguard against "shear" types of failure within the portions of the beams where it is placed. Since, therefore, the shear capacity of the portions of the shear span without transverse reinforcement corresponds to a value of the applied load significantly smaller than that leading to flexural failure, it would be expected that the load-carrying capacity of beams A and D in Fig. 1.2a and beams A1, C1, and D1 in Fig. 1.2b corresponded to shear capacity.

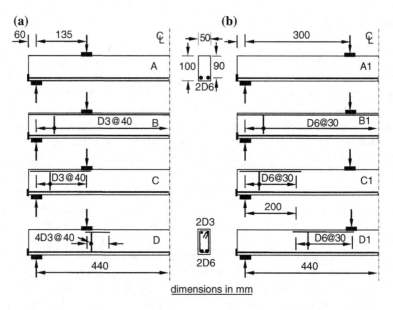

Fig. 1.2 Beams under two-point loading [21]. Design details (beams differ in the arrangement of stirrups only): **a** beams with $a_v/d \approx 1.5$; **b** beams with $a_v/d \approx 3.3$

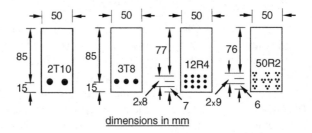

Fig. 1.3 Cross-sectional details of simply-supported beams tested under two-point loading in order to investigate the validity of the hypothesis of "dowel action" [23]; (2T10, two 10 mm dia. high-yield deformed bars; 3T8, three 8 mm dia. high-yield deformed bars; 12R4, twelve 4 mm dia. mild steel smooth bars; fifty 2 mm dia. mild smooth bars)

Yet, the experimental results depicted in Fig. 1.4 show that, in contrast with beams A and A1, which did indeed fail in "shear", beams D, C1 and D1 exhibited a flexural mode of failure. It may also be noted that the load-deflection curves of beam D and beams C1 and D1 are similar to those of beams B and C and beam B1, respectively, the latter being designed in accordance with current code provisions.

The ductility which characterises the behaviour of beam D1 is directly related to the large width of the cracks forming within the tensile zone as the beam approaches its ultimate-limit state. It is important to note that the width of the inclined crack which formed within the portion of the shear span without shear reinforcement exceeded 1 mm [21]. It has been established experimentally that such a crack width

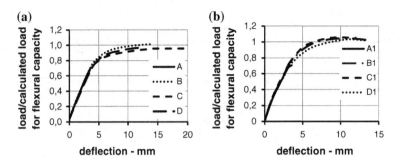

Fig. 1.4 Experimental load-deflection (of the mid cross-section) *curves* of the beams in Fig. 1.2 [21]:
a beams with $a_v/d \approx 1.5$; **b** beams with $a_v/d \approx 3.3$

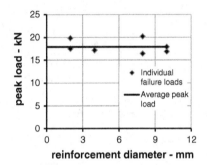

Fig. 1.5 Variation of failure load of beams in Fig. 1.3 with the size of the diameter of the
longitudinal bars [23]

precludes "aggregate interlock" even if there were a shearing movement of the crack
interfaces [24]. In conclusion, the test results clearly demonstrate that there can be
no contribution to shear capacity through "aggregate interlock" at the interfaces of
inclined cracks (a conclusion also corroborated by numerical modelling [25, 26]).

A similar conclusion is drawn for the case of "dowel action" from the results
obtained from the tests on the beams of Fig. 1.3 [23]. "Dowel action" is effected
by the bending and shear stiffness of a steel bar, and, as a result, it must be affected
by the diameter of such bars. A reduction in bar diameter should lead to a considerable reduction of the flexural and transverse stiffnesses and, hence, it is realistic to
expect a significant reduction in the contribution of "dowel action" to shear capacity. However, a reduction in the diameter of the bars used as longitudinal reinforcement for beams, such as beams A and A1 in Fig. 1.2, in a manner that maintains
the total amount of longitudinal reinforcement essentially constant (see Fig. 1.3),
was found to have no effect on the shear capacity of the beams (see Fig. 1.5) [23].

(d) Mechanism of "shear" resistance

The experimental results presented in the preceding section clearly demonstrate
that, of the auxiliary mechanisms of shear resistance, only "uncracked" concrete
in the compressive zone may contribute to the shear capacity of an RC structural

Fig. 1.6 Design details and loading arrangement of RC T-beam tested under six-point loading [27]

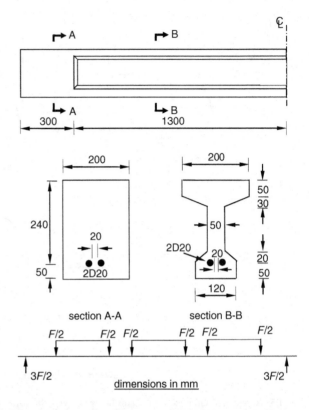

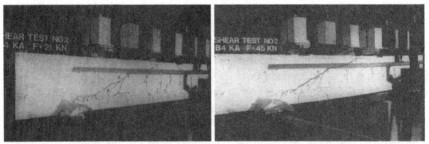

Fig. 1.7 Crack patterns of RC T-beam in Fig. 1.6 under a total load equal to 63 kN (*left*) and 135 kN (*right*) [27]

element. The causes for such behaviour may be explained by using the results obtained from tests on RC T-beams tested under six-point loading (see Fig. 1.6) [27]. Figure 1.7 shows typical crack patterns of such an RC T-beam for values of the applied load equal to 63 and 135 kN. The former of the applied values is nearly double the value predicted by current codes, while the latter is about four times larger than the code prediction of load-carrying capacity.

It is interesting to note in the figure that, in spite of the considerable increase of the applied load, the crack patterns differ only in the width of the inclined crack,

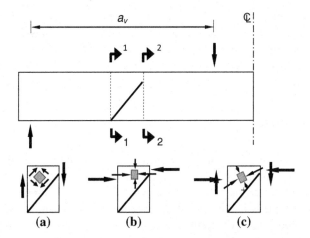

Fig. 1.8 Schematic representations of the stress conditions in the region of a deep inclined crack: **a** due to shear force; **b** due to compressive force caused by bending; and **c** due to combined action of compressive and shear force

which attained a value exceeding 3 mm for the case of the higher load [27]. As for the case of beam D1 in Fig. 1.2b, such a crack width precludes "aggregate interlock" along the crack surfaces [24]. However, the main characteristic of the crack pattern—in both cases—is the deep penetration of the inclined crack into the compressive zone which, at the cross-section including the tip of the inclined crack, has a depth of merely 10 mm. For the two values of the applied load considered above, the shear force acting at this cross-section attains values of 10.5 and 25 kN, respectively. As the size of the crack width precludes any contribution to shear capacity other than that of the compressive zone, the mean values of shear stress corresponding to the above values of shear force are 5.25 and 12.5 MPa, respectively. These values of shear stresses are indicative of the magnitude of the tensile stresses expected, in accordance with current design methods, to develop within the compressive zone in the region of the tip of the deep inclined crack. As the magnitude of the tensile stresses exceeds by a large margin the tensile strength (f_t) of concrete ($f_t \approx 0.1 \times f_c = 0.1 \times 32 = 3.2$ MPa, where $f_c = 32$ MPa the compressive strength of the concrete used for manufacturing the beams), failure should have occurred well before the lower of the values of the applied shear force considered above was attained.

However, current design methods ignore the existence of a triaxial compressive-stress field, within the region between the extreme compressive fibre and the location of the tip of the deepest inclined crack, which, as discussed later, is inevitably caused by the local volume dilation of concrete under the large longitudinal compressive stresses developing on account of bending at a cross-section with a small depth of the compressive zone [1, 20]. The existence of such a triaxial compressive stress state counteracts the tensile stresses due to the shear forces acting in the same region in the manner schematically described in Fig. 1.8; hence, the stress conditions remain compressive in this region, in spite of the presence of exceedingly high shear stresses. However, as the applied load further increases, the shear force eventually obtains a value for which the tensile stresses developing cannot be counteracted by the compressive stresses due to volume dilation and, thus, failure occurs in the manner described in Sect. 1.3.

1.2.3 Inclined Strut

The inclined struts of the truss model of an RC structural element at its ultimate-limit state forms within the web of the structural element where concrete is characterised by the presence of densely spaced inclined cracks; such cracks, for the case of cyclic/earthquake loading intersect one another as shown in Fig. 1.9 [28]. Moreover, a prerequisite for the formation of inclined struts is that concrete retains a sufficient amount of its compressive strength, after the onset of "visible" cracking, which would allow it to sustain the compressive forces assumed to be carried by the truss model's struts. As it is well known that the behaviour of "cracked" concrete is described by post-peak stress (σ)-strain (ε) characteristics [20], the above prerequisite is considered to be satisfied if concrete is characterised by strain-softening behaviour once the peak-load level is exceeded.

However, it is well known that the $\sigma - \varepsilon$ curves describing the behaviour of concrete in compression are usually obtained from tests on concrete specimens, such as, for example, cylinders or prisms, loaded through steel plates. Inevitably, therefore, the difference in the mechanical properties between concrete and steel causes the development of frictional forces at the specimen/platen interfaces. These forces restrain the lateral expansion of concrete at the end zones of the specimen, and, hence, modify the intended stress conditions in these zones.

Although one of the main objectives of current test techniques is the elimination of the above frictional forces, this objective has proved impossible to achieve to date [29]. Figure 1.10 shows characteristic stress-strain curves established from tests on cylinders in uniaxial compression by using various techniques for reducing friction at the specimen/platen interfaces [30]. From the figure, it can be seen that, in contrast with the ascending branch, which is essentially independent of the technique used

Fig. 1.9 Crack pattern of a linear RC structural element under transverse cyclic loading

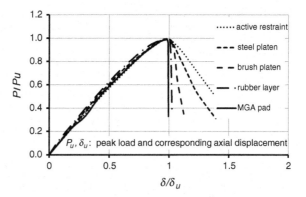

Fig. 1.10 Axial stress-axial strain *curves* obtained from uniaxial-compression tests on concrete cylinders using various means to reduce friction at the specimen-loading platen interfaces [30]

to reduce friction, the slope of the descending branch increases with the efficiency of the friction-reducing medium employed. In fact, the increase in slope is such that it leads to the conclusion that, if it were possible to eliminate friction entirely the descending branch would have a 90° slope, which is indicative of an immediate and complete loss of load-carrying capacity as soon as the peak stress is attained.

It appears from the above, therefore, that the "softening" branch of a stress-strain curve essentially describes the interaction between specimen and loading platens and *not*, as widely considered concrete behaviour. Concrete behaviour is described only by the ascending branch of an experimentally established $\sigma - \varepsilon$ curve, and loss of load-carrying capacity occurs in a brittle manner. (A similar conclusion may also be drawn from the experimental information presented in Ref. [31], as well as that obtained from an international co-operative project organised by RILEM TC-148SSC [29].) Due to its brittle mode of failure, therefore, concrete does not have sufficient residual strength that would allow the formation of inclined cracks within an RC structural element's web.

1.3 Flexural Capacity

Amongst the assumptions underlying the assessment of flexural capacity is that the behaviour of concrete in the compressive zone is adequately described by $\sigma - \varepsilon$ curves, comprising both an ascending and a gradually descending branch, established from tests on cylinders or prisms in uniaxial compression. This assumption, on the one hand attributes the strains, of the order of 0.35 %, measured at the extreme compressive fibre of an RC beam at its ultimate-limit state in flexure, to the strain-softening behaviour of concrete, and, on the other hand, implies that the effect of small transverse stresses, which invariably develop in any RC structural element, on concrete behaviour is insignificant.

And yet, this assumption is not valid on both counts: as discussed in the preceding section, concrete is, in nature, a brittle, rather than a softening, material, whereas the small transverse stresses have a considerable effect on concrete

Fig. 1.11 Strength envelope
of concrete of concrete under
axisymmetric states of stress

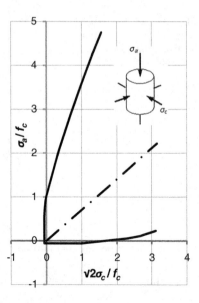

Fig. 1.12 RC beams under
two-point loading: design
details [32]

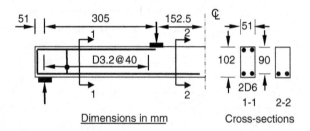

behaviour [20]. In fact, Fig. 1.11 shows that a transverse stress of the order of 0.1
f_c may either increase load-carryin-g capacity by over 50 %, when compressive, or
reduce it to zero, when tensile.

Further to the discussion of the post-peak behaviour of concrete in Sect. 1.2.3, the
irrelevance of strain softening to structural-concrete behaviour may be demonstrated
by reference to the results obtained from tests on three simply-supported RC beams,
with a rectangular cross-section, subjected to two-point loading [32]. The details
of a typical beam are shown in Fig. 1.12, with the central portion in pure flexure
constituting one-third of the span. Besides the load measurement, the deformational
response was recorded by using 20 mm long electrical resistance strain gauges and
linear-voltage differential transducers (LVDTs). The strain gauges were placed on
the top and side surfaces of the beams in the longitudinal and transverse directions as
shown in Fig. 1.13. The figure also indicates the position of the LVDTs which were
used to measure deflection at mid-span and at the loaded cross-sections.

Of the results obtained from the above tests, Fig. 1.14 shows the relationships between longitudinal (i.e. along the beam axis) and transverse (i.e. across

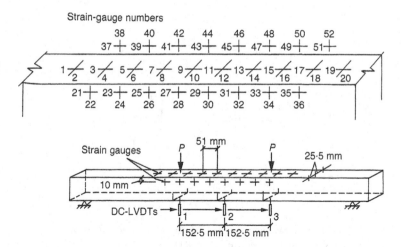

Fig. 1.13 RC beams under two-point loading: beam instrumentation [32]

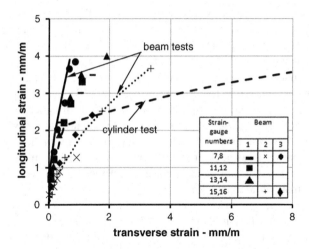

Fig. 1.14 Relationships between longitudinal and transverse strains measured on the *top* surface between the load points of the RC beams in Fig. 1.12 (for strain gauge locations see Fig. 1.13) [32]

the beam width) strains, as measured on the top surface of the girders, throughout the middle third of the beam span. Also plotted in the figure is the relationship between longitudinal and transverse strains derived from $\sigma - \varepsilon$ relationships (shown in Fig. 1.15) established from tests on cylinders under uniaxial compression [1]. Now, if the relationships of Fig. 1.15 were to provide a realistic description of concrete behaviour in the compressive zone of the beams tested in flexure, then one would expect the relationships between longitudinal and transverse strains measured on the top surface of the beams to be compatible with their counterparts established from the cylinder test. Furthermore, longitudinal

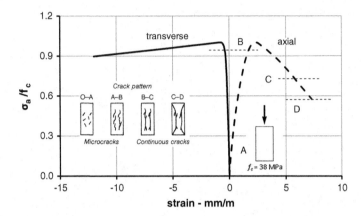

Fig. 1.15 RC beams under two-point loading: stress–strain relationships and crack patterns under uniaxial compression for the concrete mix used [1]

macrocracks ought to appear on the top surface of the beams, as indicated in Fig. 1.15, where typical crack patterns of axially-compressed concrete cylinders around (B–C) and beyond (C–D) the peak-load level are depicted schematically. It is apparent from Fig. 1.14, however, that, for the region of the cross-section including a primary flexural crack, only the portion of the deformational relationship based on the uniaxial cylinder test up to a level (B) close to the peak-load level can provide a realistic description of the behaviour of concrete in the compressive zone of the beam. Beyond this level, there is a dramatic deviation of the cylinder strains from the beam relationships. Not only does such behaviour support the view that the post-peak branch of the deformational response of the cylinder in compression does not describe material response but, more importantly, it clearly proves that, while uniaxial $\sigma - \varepsilon$ data may be useful prior to the attainment of the peak stress, they are insufficient to describe the behaviour of concrete in the compressive zone once this maximum-stress level is approached.

An indication of the causes of behaviour described by the relationships of Fig. 1.14 may be seen by reference to Fig. 1.16, which shows the change in shape of the transverse deformation profile of the top surface of one of the beams (but typical for all beams tested) with load increasing to failure [32]. The characteristic feature of these profiles is that, within the 'critical' central portion of the beam, they exhibit large local tensile strain concentrations which develop in the compressive regions of the cross-sections where primary flexural cracks, that eventually cause collapse, occur. Such a large and sudden increase in transverse expansion near the ultimate load is indicative of volume dilation and shows quite clearly that, even in the absence of stirrups, a triaxial state of stress can be developed in localised regions within the compressive zone. The local transverse expansion is restrained by concrete in adjacent regions (as indicated by the resultant compression forces F in Fig. 1.16), a restraint that has been found to be equivalent to at least 10 % of f_c [32]; hence, as Fig. 1.11 indicates, the compressive region in the

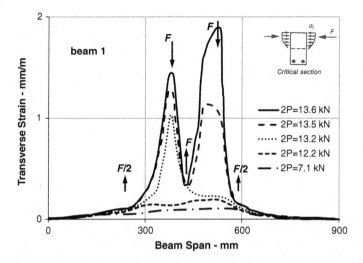

Fig. 1.16 Typical variation of the deformation profile of the loaded face of the RC beams in Fig. 1.12 with increasing total load (2P) and schematic representation of resulting forces (F) and stresses (σ_t) [32]

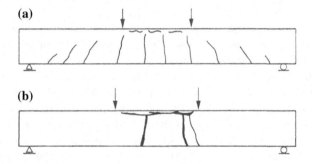

Fig. 1.17 Typical failure mode of RC beams in flexure: **a** a schematic representation of crack pattern at collapse; **b** observed failure of test beams following collapse [32]

plane of the main flexural crack is afforded a considerable increase in strength so that failure does not initiate there. Concurrently, the expanding concrete induces tensile stresses in the adjacent regions (these are indicated by the resultant tension forces F and $F/2$ in Fig. 1.16), and this gives rise to a compression/tension state of stress. Such a stress state reduces the strength of concrete in the longitudinal direction, and collapse occurs as a result of horizontal splitting of the compressive zone in regions between primary flexural cracks, as illustrated schematically in Fig. 1.17. Concrete crushing, which is widely considered to be the cause of flexural failure, thus appears to be a *post-failure* phenomenon that occurs in the compression zone of cross-sections containing a primary flexural crack due to loss of restraint previously provided by the adjacent concrete.

It may be concluded from the above, therefore, that the large compressive and tensile strains measured on the top surface of the central portion of the beams should be attributed to a *multiaxial* rather than a uniaxial state of stress. A further indication that these large strains cannot be due to post-ultimate $\sigma - \varepsilon$ characteristics is the lack of any *visible* longitudinal cracking on the top surface for load levels even near the load-carrying capacity of the beams. As shown in Fig. 1.15, such cracks characterise the post-ultimate strength behaviour of concrete under compressive states of stress.

1.4 Critical Regions

In regions—referred to in Codes [3, 4] as *critical* regions—of linear RC elements (such as, for example, beams and columns) where a large bending moment and a large shear force develop concurrently, current code provisions for earthquake-resistant design specify an amount of stirrup reinforcement significantly larger than that safe-guarding against shear types of failure. This additional stirrup reinforcement is placed in order to provide confinement to concrete within the compressive zone, which restrains its lateral expansion and increases its strength and ductility in the longitudinal direction, thus leading to a significant improvement of the ductility of the RC member.

However, there has been published experimental evidence [6, 9], in recent years, obtained from tests on beam/column elements exhibiting points of contraflexure, which shows that there are cases for which the additional amount of transverse reinforcement may cause a brittle type of failure, rather than safeguard ductile structural behaviour. Such types of failure, which are characterised by the presence of inclined cracks penetrating deeply into the compressive zone of the critical regions, are indicated in Fig. 1.18 (top) and (bottom). The former of these figures shows the mode of failure of a simply-supported beam with an overhang, reinforced in compliance with EC2/EC8, which was subjected to sequential point loading; a point load was first applied at mid span and increased to a value close, but, not beyond, the beam flexural capacity, where it was maintained constant while a second point load was applied in the overhang and increased monotonically to failure [9]. The latter figure shows the mode of failure of the critical region of the portion of a two-span linear element, also designed in compliance with EC2/EC8, modelling to a 1:3 scale a column between consecutive floor levels [6]. This element was subjected to the action of a constant axial load combined with lateral cyclic loading.

The causes of the above brittle types of failure are considered to relate to the experimental information used by the code methods for assessing the transverse reinforcement required for the critical regions of RC beam-like elements. This experimental information was obtained from uniaxial-compression tests on concrete cylinders or prisms subject to lateral confinement through the use of spiral or stirrup reinforcement. And yet, unlike the cylinders or prisms which are subjected to uni-axial compression, the critical regions of beam-like elements are subject to the combined action of a bending moment and a shear force which causes the formation of inclined cracks. Such cracks penetrate deeply into the compressive zone and have the tendency to extend near horizontally (in the direction of the maximum principal

Fig. 1.18 Failure of the critical region of a simply-supported RC beam with overhang exhibiting point of contraflexure (*top*) [9], and a two-span linear RC element under the action of a constant axial force combined with cyclic lateral loading (*bottom*) [6]

compressive stress) due to the presence of large, near vertical tensile stress concentrations which develop in the region of the crack tips [18, 19]. The magnitude of these tensile stresses is such that their resultant may eliminate the vertical component of the confinement considered to be provided by the stirrups, and, therefore, the expansion of concrete in the vertical direction may remain essentially unaffected by the stirrups. Moreover, the presence of significant inclined cracking reduces the strength of concrete within the element's web, and this reduction, combined with the larger transverse compression applied to concrete by the excess amount of stirrup reinforcement anchored to it, may lead to premature failure within the critical regions [1].

1.5 Points of Contraflexure

During the 7/9/1999 Athens earthquake many reinforced concrete (RC) structures [particularly those lacking symmetric plan configuration and containing a "soft" ground-floor storey (pilotis)] suffered unexpected brittle damage that cannot be attributed to either non-compliance with code provisions or defective work [5]. Examples of this damage are presented in Fig. 1.19, which shows the unexpected failure suffered by vertical structural members at the location of the point of contraflexure usually situated within the mid-height region of the member. This type of failure, which has been reproduced under controlled laboratory conditions, is not taken into consideration by the methods adopted by current codes of practice for the design of RC structures (invariably based on the truss analogy (TA) [10, 11]) [2–4].

The relevant feature of the modes of failure shown in Fig. 1.19 is not the occurrence of criss-crossing diagonal cracking, but the location of the region where cracking occurred. In all cases, the location of failure was found to lie within the region of the point of contraflexure. Such an event could not be simply attributed to coincidence, since, as discussed above, this location of failure repeatedly characterises a large number of structural elements that suffered the above type of damage during the 7/9/99 earthquake in Athens.

It appears realistic to seek the causes of the above mode of failure in the form that the truss model takes in the region of the point of contraflexure. Figure 1.20

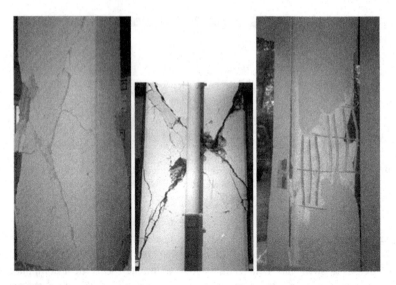

Fig. 1.19 Examples of damage suffered by columns of RC structures during the 1999 Athens earthquake [5]

depicts a truss modelling a vertical linear structural element subjected at its end-faces to the combined action of an axial force (N), a bending moment (M) and a shear force (V). The figure shows that the vertical element is modelled by means of two antisymmetric trusses, connected in the region of the location of the point of inflection with a transverse tie. The shaded portions of the trusses represent the flow of the compressive stresses from the cross-sections with zero bending moment to the cross-sections with a near-constant depth of compressive zone.

The above model implies that in the region of the point of inflection concrete is locally subjected to a *direct transverse tensile*, rather than shear, force causing failure when the tensile strength of the material is exceeded. This type of failure may be prevented by specifying stirrups, forming the transverse tie indicated in Fig. 1.20, in a quantity sufficient to sustain the tensile force in excess of that which can be sustained by concrete alone, and this being a simple strength requirement. Here, it is essential to appreciate that contraflexure (i.e. points of inflection) is associated with the kind of response exhibited by a beam-column in a building subjected to lateral sway (but applicable to any beam or frame with a point of inflection) as shown in Fig. 1.21: it is evident that the tie is needed to prevent separation of the two ends of the constituent members.

Now, designing the reinforcement in the region of the point of inflection for shear, rather than direct tension, leads to a considerable underestimate of the quantity of stirrups required to prevent failure. This is because the provisions of the code adopted for shear design allow, not only for the contribution to shear resistance of the auxiliary mechanisms discussed in Sect. 1.2.2, but, also, for the beneficial effect of the axial force. And yet, within an essentially tensile stress field the above auxiliary mechanisms cannot develop, whereas it is well established that the presence of an axial force is likely to reduce the tensile strength of concrete in the orthogonal direction.

Fig. 1.20 Truss modelling a column subjected to the combined action of axial force (N), bending moment (M) and shear force (V) at its end faces

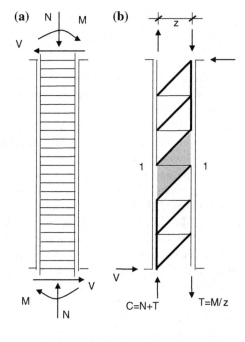

Fig. 1.21 Beam-column exhibiting a point of inflection in a structure (*left*) and illustration of an internal tie needed at contraflexure in order to prevent separation of the two ends of the constituent members

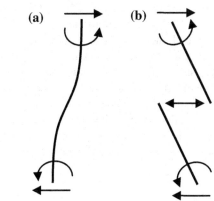

1.6   Effects on Structural Behaviour

From the discussion in the preceding sections, it becomes clear that the conflict between the concepts underlying current code design methods and the causes of structural behaviour mainly relates with the ultimate limit state of a structure or structural member. It should be expected, therefore, that were there any effects of this conflict on the observed behaviour of a structural element designed by using code methods, these should be sought in situations where the structural element reaches its ultimate limit state. In real structures, such situations may arise in cases of overload; alternatively, they may be reproduced through the testing of model elements under controlled laboratory conditions.

Although shortcomings of current-code methods such as those described in the preceding section may, to some extent, be counteracted by implementing the code requirements for nominal reinforcement and detailing, there have been some notable structural failures that may have been prevented, had the design methods been based on sound concepts providing a realistic description of concrete behaviour as a material. One such failure is the collapse of the "Sleipner 4" platform in the North Sea at a depth of 40 m, during the sinking operation for positioning it on the seabed at a depth of some 140 m; this has been blamed on the inadequacy of the ACI shear-design provisions [33]. Another has been the collapse of a multilevel car park, in Wolverhampton, U.K., which occurred due to punching of the top level flat slab under dead load only [34]. It has been reported that such premature punching was preceded by loss of bond of the longitudinal reinforcement [35]; loss of bond may lead to a brittle type of failure in the manner discussed in Chap. 2.

Unexpected types of brittle failure often occur in earthquake-stricken regions; typical is the case (discussed in Sect. 1.5) of the significant damage suffered by the vertical elements of RC buildings in the region of points of inflection during the 1999 Athens earthquake [5]. Such damage occurred not only in structures designed to old code provisions, but, also, in structures satisfying the performance requirements of current codes which are widely deemed to safeguard ductile behaviour. In fact, it is interesting to note in Fig. 1.19 that the damage suffered by the "slender" column in Fig. 1.19 (left) was similar not only to that of the "short" column in Fig. 1.19 (middle) (both these columns had been designed to current code provisions), but, also, to that of the "slender" column in Fig. 1.19 (right) which had been designed to older, less stringent code provisions based on the permissible-stress philosophy [5]. This type of damage is a typical example of damage reflecting the lack of a sound theory underlying the methods adopted by current codes for the design of RC structures. Serious doubts regarding the validity of the concepts underlying earthquake-resistant design have already been expressed elsewhere [36–38], whereas a thorough account of failures suffered by RC structures has been the subject of other publications [39].

1.7 Alternative Design Methods

It appears from the above that the time is ripe for considering radical changes in RC design involving the replacement of the concepts underlying the methods currently adopted by codes with new ones compatible with structural concrete behaviour. Such radical changes have already been attempted [1, 40, 41]; as early as the mid-sixties, the shortcomings of the methods adopted by the codes were attributed to the criteria used for the prediction of brittle types of failure [40], which code provisions have always linked with the shear capacity of RC structural elements. In fact, it has been proposed that a more realistic criterion should treat brittle failure as a premature loss of flexural capacity due to the combined action of bending moment and shear force [40]. Not only has such a failure criterion [41] (in a form suitable for practical applications) been developed and incorporated in a design guidance report [42], but, also, it has been employed in practice for the design of original structures such as, for example, the Calgary coliseum [43], the grandstand roof of the rugby stadium at Twickenham, etc. More recently, combined with concepts that allow for a realistic description of concrete behaviour (as established from comprehensive investigations of the fundamental characteristics of the deformational response and failure mechanism of concrete at both the material and structure levels), the above failure criterion led to the development of a unified design method—the method of the compressive-force path—found to consistently satisfy the performance requirements of current codes [1, 20].

Design methods such as those described in Refs. [1, 41, 42] are considered to point towards the orientation that should be given to research efforts, if such efforts are to lead to a significant improvement of the codes of practice for RC design.

1.8   Concluding Remarks

The concepts which form the basis of current codes of practice for the design of RC structures are in conflict with fundamental properties of concrete at both the material and the structure levels.

This conflict is reflected on the premature brittle types of failure unexpectedly suffered by RC structures in situations of overload.

Such types of failure, which have been reproduced under controlled laboratory conditions, may be prevented through the use of alternative design methods that allow for "true" structural-concrete behaviour.

Already published work aiming to developing alternative design methods points towards the type of research required for achieving a significant improvement of the provisions of current codes of practice for the design of RC structures.

References

1. Kotsovos MD, Pavlovic MN (1999) Ultimate limit-state design of concrete structures: a new approach. Thomas Telford (London), p 164
2. American Concrete Institute (2002) Building code requirements for structural concrete (ACI 318-02) and commentary (ACI 318R-02)
3. EN 1992-1, Eurocode 2 (2004) Design of concrete structures—part 1-1: general rules and rules for buildings
4. EN 1998-1, Eurocode 8 (2004) Design of structures for earthquake resistance—part 1: general rules, seismic actions and rules for buildings
5. Kotsovos MD, Pavlovic MN (2001) The 7/9/99 Athens earthquake: causes of damage not predicted by structural-concrete design methods. J Struct Eng 79(15):23–29
6. Kotsovos MD, Baka A, Vougioukas E (2003) Earthquake-resistant design of reinforced-concrete structures: shortcomings of current methods. ACI Struct J 100(1):11–18

7. Kotsovos GM, Zeris C, Pavlovic MN (2005) Improving RC seismic design through the CFP method. In: Proceedings of ICE, buildings and structures, vol 158(SB5), pp 291–302
8. Kotsovos GM, Zeris C, Pavlovic MN (2007) Earthquake-resistant design of indeterminate reinforced-concrete slender column elements. Eng Struct 29(2):163–175
9. Jelic I, Pavlovic MN, Kotsovos MD (2004) Performance of structural-concrete members under sequential loading and exhibiting points of inflection. Comput Concr 1(1):99–113
10. Ritter W (1899) Die Bauweise Hennebique. Schweisserische Bauzeitung, vol 33, pp 59–61
11. Morsch E (1902) Versuche uber Schubspannungen in Betoneisentragen. Beton und Eisen, Berlin, vol 2(4), pp 269–274
12. Barnard PR (1964) Researches into the complete stress-strain curve for concrete. Mag Concr Res 16(49):203–210
13. Fenwick RC, Paulay T (1968) Mechanisms of shear resistance of concrete beams. J Struct Div. ASCE proceedings, vol 94(ST10), pp 2325–2350
14. Taylor HPJ (1974) The fundamental behaviour of reinforced concrete beams in bending and shear. Shear in reinforced concrete, ACI publication SP-42, American concrete institute, pp 43–77
15. Taylor HPJ (1969) Investigation of the dowel shear forces carried by tensile steel in reinforced concrete beams. Technical report 431 (publication 42.431), cement and concrete association, London
16. Collins MP, Mitchell D (1980) Shear and torsion design of prestressed and non-prestressed concrete beams. Prestressed Concr Inst 25(5):32–100
17. Schlaich J, Schafer K, Jennewein M (1987) Toward a consistent design of structural concrete. Prestressed Concr Inst 32(3):74–150
18. Kotsovos MD (1979) Fracture of concrete under generalised stress. Mater Struct RILEM 12(72):151–158
19. Kotsovos MD, Newman JB (1981) Fracture mechanics and concrete behaviour. Mag Concr Res 33(115):103–112
20. Kotsovos MD, Pavlovic MN (1995) Structural concrete: finite-element analysis for limit-state design. Thomas Telford, London, p 550
21. Kotsovos MD (1987) Shear failure of reinforced concrete beams. Eng Struct 9(1):32–38
22. Kotsovos MD (1987) Shear failure of RC beams: a reappraisal of current concepts. CEB Bull 178/179:103–111
23. Jelic I, Pavlovic MN, Kotsovos MD (1999) A study of dowel action in reinforced concrete beams. Mag Concr Res 51(2):131–141
24. Reinhardt HW, Walraven JC (1982) Cracks in concrete subject to shear. J Struct Div, proceedings of the ASCE 108(ST1):207–224
25. Kotsovos MD (1984) Behaviour of reinforced concrete beams with a shear span to depth ratio between 1.0 and 2.5. ACI J. Proceedings 81(3):279–286. May–June 1984
26. Kotsovos MD (1986) Behaviour of RC beams with shear span to depth ratios greater than 2.5. ACI J. Proceedings 83(115):1026–1034. Nov–Dec 1986
27. Kotsovos MD, Bobrowski J, Eibl J (1987) Behaviour of RC T-beams in shear. Struct Eng 65B(1):1–9
28. Kotsovos G (2005) Improving RC seismic design through the CFP method. In: Proceedings of the institution of civil engineers, structures and buildings, vol 158(SB5), pp 291–302. Oct 2005
29. van Mier JGM, Shah SP, Arnaud M, Balayssac JP, Bascoul A, Choi S, Dasenbrock D, Ferrara G, French C, Gobbi ME, Karihaloo BL, Konig G, Kotsovos MD, Labuz J, Lange-Kornbak D, Markeset G, Pavlovic MN, Simsch G, Thienel K-C, Turatsinze A, Ulmer U, van Geel HJGM, van Vliet MRA, Zissopoulos D (1997) Strain-softening of concrete in uniaxial compression. Mater Struct RILEM 30(198):195–209. (Report of the round robin test carried out by RILEM TC 198-SSC: test methods for the strain-softening response of concrete.)
30. Kotsovos MD (1983) Effect of testing techniques on the post-ultimate behaviour of concrete in compression. Mater Struct RILEM 16(91):3–12
31. Van Mier JGM (1986) Multiaxial strain-softening of concrete. Mater Struct RILEM 19(111):179–200

32. Kotsovos MD (1982) A fundamental explanation of the behaviour of reinforced concrete beams in flexure based on the properties of concrete under multiaxial stress. Mater Struct RILEM 15(90):529–537
33. Collins MP, Vecchio FJ, Selby RG, Gupta PR (1997) The failure of an offshore platform. Concr Int 28–34. Aug 1997
34. Shock collapse sparks lift slab fears and safety experts urge car park review (1997) New civil engineer, pp 3–4. 27 Mar/3 April 1997
35. Kellermann JF (1997) Riper row car park, Wolverhampton: results of the investigation. Conference on concrete car parks: design and maintenance issues held at the Cavendish Centre, London, 29 Sept 1997, British Cement Association
36. Priestley MJN (1997) Myths and fallacies in earthquake engineering: conflicts between design and reality. Concr Int 54–63. Feb 1997
37. Hansford M (2002) Seismic codes oversimplified and unsafe. New civil engineer, 8/15 Aug 2002, p 28. (Report on seminar by V. Bertero organised jointly by ICE society for earthquake and civil engineering dynamics and Wessex institute of technology
38. Priestley MJN Revisiting myths and fallacies in earthquake engineering, the ninth Mallet-Milne lecture organised by the society for earthquake and civil engineering dynamics
39. Carpaer KL (1998) Current structural safety topics in North America. Struct Eng 76(12):233–239
40. Kani GNJ (1964) The riddle of shear and its solution. J Am Concr Inst Proc 61(4):441–467
41. Bobrowski J, Bardham-Roy BK (1969) A method of calculating the ultimate strength of reinforced and prestressed concrete beams in combined flexure and shear. Struct Eng 47(5):197–209
42. The Institution of Structural Engineers (1978) Design and detailing of concrete structures for fire resistance, interim guidance by a joint committee of the institution of structural engineers and the concrete society, April 1978, p 59
43. Il≪Saddledome≫: stadio olimpico del ghiaccio a Calgary (Canada) (1984) Progetto struttur-ale: Jan Bobrowski and Partners Ltd. Progetto architettonico: Graham McCourt. *L' Industria Italiana del* Cemento. No. 5

Chapter 2
The Concept of the Compressive-Force Path

2.1 Introduction

This chapter presents a qualitative description of the behaviour and function of a structural concrete member at its ultimate limit state, together with a description of the mechanism which underlies the transfer of external load from its point of application to the supports of the structural member. This qualitative description, which is compatible with all available experimental information, is made by reference to the case of a simply-supported beam, without stirrups, at its ultimate limit state under transverse loading (the effect of axial loading is also considered). Such a structural member is chosen because, not only is there ample experimental information describing its behaviour but, also, the description of how the beam actually functions forms the theory underlying the design methodology proposed in the following chapters. This theory has been termed the 'compressive-force path (CFP) concept' since, as deduced from the description of how the beam functions, the main characteristic of the beam is that both its loading capacity and failure mechanism are related to the region of the member containing the path of the compressive stress resultant which develops within the beam due to bending, just before failure occurs. Experimental information on the validity of the concept is also presented, and it is shown that this provides a realistic description of the causes which dictate the various types of beam behaviour as established by the experimental information available to date. The generalisation of the concept, so as to extend its applicability to any structural configuration and, in particular, to the case of frame-type structures, forms the subject of the Chap. 6.

2.2 Proposed Function of Simply-Supported Beams

2.2.1 Physical State of Beam

Figure 2.1 provides a schematic representation of the crack pattern and the deflected shape (in a magnified form) of a simply-supported beam under transverse loading, just before failure. The figure shows that cracking encompasses a

M. D. Kotsovos, *Compressive Force-Path Method*, Engineering Materials,
DOI: 10.1007/978-3-319-00488-4_2, © Springer International Publishing Switzerland 2014

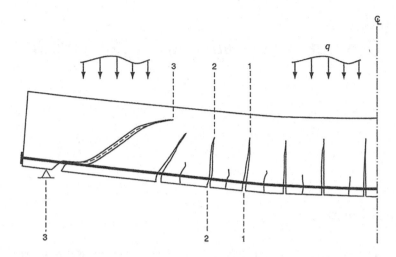

Fig. 2.1 Schematic representation of crack pattern and deformed shape of a simply-supported beam under transverse load

large portion of the beam and comprises both vertical and inclined cracks. The cracks, in most cases, initiate at the bottom face of the beam and, having propagated through the beam web, penetrate deeply into the compressive zone, the crack tip moving closer to the upper face. As will be seen in Sect. 2.2.3, when the causes of failure are associated with the presence of the deep inclined crack closest to the support, this crack not only penetrates into the compressive zone deeper than any other crack, but also extends towards the support along the longitudinal tension bars, destroying the bond between the bars and the surrounding concrete.

It would appear from Fig. 2.1, therefore, that concrete eventually remains uncracked only within a relatively small portion of the beam. This portion includes, on the one hand, the two end regions of the beam which extend to the deep inclined crack forming closest to the supports and, on the other hand, the relatively narrow strip, with varying depth, which forms between the crack tips and the upper face, and connects the above two regions. As will become apparent in what follows, a characteristic feature of the above narrow strip is its very small depth which, as indicated in the figure, is, in localized regions (and, in particular, in the region including the tip of the deepest inclined crack), a very small percentage of the total beam depth.

It should be noted that the presence of external load acting on the end faces of the beam, in the axial direction, may have the following two effects on the physical state of the beam depicted in Fig. 2.1.

(a) The depth of the horizontal uncracked zone of the beam may increase or decrease (leading to a corresponding reduction or increase in the length of the flexural cracks), depending on whether the axial force is compressive or tensile respectively.
(b) The presence of an axial force may prevent the formation of any deep inclined crack.

2.2.2 Load Transfer to Supports

In spite of the extensive cracking, the beam at its ultimate limit state is capable of fulfilling its purpose, i.e. transferring the applied load to the supports. The mechanism through which this transfer is effected can only be in the form of 'beam action' adjusted so as to allow for the particular characteristics of RC members.

In any cross-section (in which the presence of an external axial load is ignored for purposes of simplicity), internal actions may be resolved into axial and transverse components. In particular, for the case of a cross-section including a deep flexural crack (such as, for example, cross-section 2-2 in Fig. 2.1), the axial internal actions are such that their combined action is equivalent to the bending moment which develops in this cross-section as a result of the external load, while the shear force is equivalent to the resultant of the external transverse forces acting on the beam portion to the left of the cross-section in question (see Fig. 2.2).

The relationship between the internal axial and shear forces may be derived by considering the equilibrium conditions of an element of the beam between two cross-sections including consecutive flexural cracks such as, for example, the element between sections 1-1 and 2-2 in Fig. 2.1, which is also illustrated in isolation as a free body in Fig. 2.3. The action of the couple arising from the shear forces that develop at the two end cross-sections of this element equilibrates the change in the bending moment between the two cross-sections. This change of bending moment is predominantly due to change in the magnitude of the axial internal actions, i.e. the compressive force sustained by concrete and the—numerically equal to it (for purposes of equilibrium)—tensile force sustained by the longitudinal steel bars (see Fig. 2.3b).

A necessary prerequisite for the change in magnitude of the above longitudinal internal actions is the existence of bond between concrete and steel, through which a portion (ΔT) of the tensile force acting on the steel bars is transferred to the concrete (see Fig. 2.3c). It should be noted that the force ΔT is the *only* action

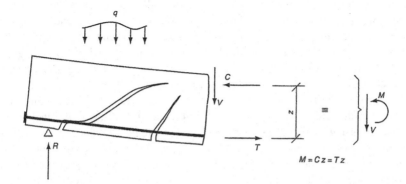

Fig. 2.2 Internal actions equivalent to the bending moment and shear force acting at a cross-section including a crack

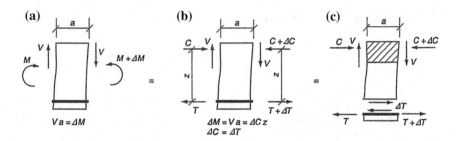

Fig. 2.3 Portion of beam (in Fig. 2.1) between two cross-sections including consecutive cracks with internal actions in (**a**) being equivalent to those in (**b**) and (**c**)

developing on any of the concrete strips between consecutive flexural or inclined cracks, since the evidence presented in Sect. 1.2.2 precludes the development of any significant forces at the crack surfaces due to 'aggregate interlock', while 'dowel action', even if it were to develop, is insignificant.

A concrete strip such as the above may be considered to function as a 'cantilever' fixed on the compressive zone of the beam and subjected o the action ΔT transmitted from the steel to concrete through bond (see Fig. 2.3c) [1]. The bending moment that develops at the cantilever base, owing to the above force, balances the action arising from the couple of the shear forces which act in the compressive zone of the beam. In fact, the above equilibrium condition essentially describes the mechanism through which the external load, in the form of shear forces, is transferred throughout the length of the span within which bond develops between concrete and steel (see Fig. 2.4a).

However, as discussed in Sect. 2.2.1, the existence of the deep inclined crack near the beam support causes bond failure, the latter extending between the intersection of the inclined crack with the longitudinal steel bars and the support, and thus the external load cannot be transferred by 'cantilever bending' beyond the section which includes the tip of the inclined crack. The mechanism through which the external load is transferred from the above section to the support becomes apparent by considering the equilibrium conditions of the end portion of the beam which encompasses the region enclosed by the end, upper, and lower faces of the beam, the inclined crack closest to the support, and the cross-section through the tip of this crack. This portion is isolated from the beam and represented schematically by the free body illustrated in Fig. 2.4b.

Owing to the destruction of the bond between concrete and the longitudinal bars, the tensile force sustained by the reinforcement is transmitted unchanged from the right-hand side of the free body to the region of the support where it combines with the reaction to yield the compressive force C'_i (see Fig. 2.4b). Similarly, the shear and axial compressive forces acting on the upper part of the right-hand side end of the free body combine to form the inclined compressive force C_i (see also Fig. 2.4b) which, for equilibrium purposes, must fulfil the condition $C_i = C'_i$. This condition indicates that it is through the development of the

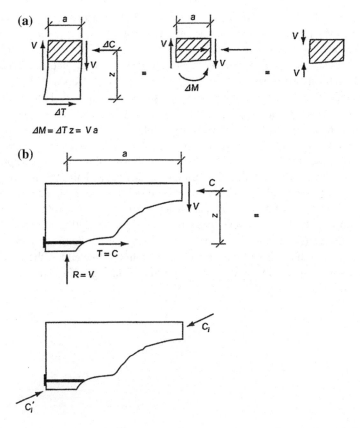

Fig. 2.4 Mechanisms of external-load transfer to the supports: **a** cantilever action, and **b** change in direction of compressive force

inclined compressive force C_i that the external load is transferred from the right-hand side of the free body to the support. In fact, the development of the above inclined force essentially represents a change in the direction of the path of the near-horizontal (within the middle portion of the beam's) compressive stress resultant which develops on account of the bending of the beam, with the change in the path direction occurring in the region of the tip of the inclined crack closest to the support.

2.2.3 Effect of Cracking on Internal Actions

An indication of the internal state of stress and the magnitude of the stresses which develop in cracked concrete may also be obtained by considering the forces acting on the beam element illustrated in Fig. 2.3. This element lies between two cross-sections (1-1 and 2-2 in Fig. 2.1) which include consecutive cracks and, as discussed in the preceding section, the only action exerted on the tensile zone of the element is the portion of the tensile force (ΔT in Fig. 2.3c) which is transferred from the longitudinal steel bars to concrete through bond. In fact, the crack surfaces, which form boundaries to this element remain stress-free since, as discussed in the preceding section, the cracking mechanism of the beam precludes the development of both 'aggregate interlock' and 'dowel action' which are the most likely mechanisms that could allow for the development of forces at the crack faces.

As discussed in the preceding section, therefore, the portion of this element between the cracks acts as a plain-concrete cantilever (fixed to the compressive zone of the beam) which undergoes bending as a result of the tensile force ΔT transmitted from the steel bars to concrete through bond (see Fig. 2.5a). The state of stress which is compatible with 'cantilever bending' results from the development of, on the one hand, a shear force constant throughout the cantilever length and equal to ΔT (see Fig. 2.5c), and, on the other hand, a bending moment, the magnitude of which increases with the distance from the free end of the cantilever, attaining its maximum value at the cross section (3-3 in Fig. 2.5a) which coincides with the fixed end.

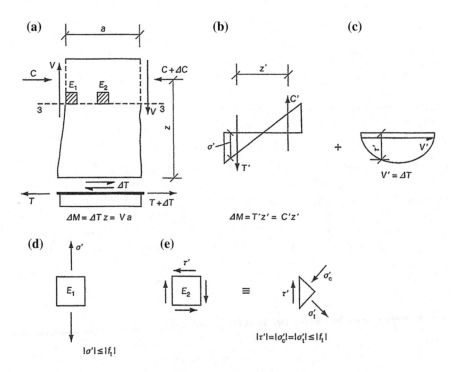

Fig. 2.5 Internal forces on portion of beam depicted in Fig. 2.3c (**a**), with normal and shear stress distributions and corresponding stress resultants being indicated in (**b**) and (**c**), respectively, whereas the stress states of elements E1 and E2 are shown in (**d**) and (**e**)

Figure 2.5b and c show the distributions of the normal (σ') and shear (τ') stresses and the corresponding stress resultants ($T' = C'$, V'), whose combined action ($T'z' = C'z'$, V') is, for purposes of equilibrium, equivalent to that of the bending moment [$\Delta M = \Delta Tz = Va$ (see Fig. 2.5a)] and shear force ($V' = \Delta T$) acting on the cross-section (3-3 in Fig. 2.5a) coinciding with the fixed end of the cantilever. Since the cantilever consists of plain concrete, the (numerically) maximum value of the stresses developing at the above cross-section cannot exceed the strength (f_t) of concrete in tension (as indicated in Fig. 2.5d and e which depict the stress conditions at two typical elements E1 and E2 (of cross-section 3-3 in Fig. 2.5a in pure tension and pure shear respectively). (For such stress values, concrete behaviour is essentially linear and, hence, the shape of the stress distributions assumed in Fig. 2.5b and c is that predicted by the simplified beam theory.)

As the fixed end of the cantilever essentially coincides with the interface between the uncracked (compressive) and cracked (tensile) zones of the beam, the uncracked zone is also subjected to the internal stresses and stress resultants acting at this interface, as indicated in Fig. 2.5b and c. However, the main actions that develop within uncracked concrete are those indicated in Fig. 2.2 which, together with the tensile force sustained by the longitudinal reinforcement, resist the combined action of the bending moment and shear force caused by the applied load. Figure 2.2 indicates that uncracked concrete (i.e. the compressive zone of the beam cross-section 2-2 in Fig. 2.1) is subjected not only to the axial compressive force C (due to the bending moment) but also to the *total* shear force acting at the beam cross-section (since, as discussed earlier, cracked concrete cannot contribute to the shear resistance of the beam). Here, it should be recalled that, although the magnitude of the nominal shear stress (i.e. the ratio of the shear force to the area of the compressive zone of the cross-section) exceeds (in regions where the depth of the uncracked concrete is small) the concrete shear capacity (as defined in current codes) by a large margin, the mechanism of shear resistance described in Sect. 1.2.2(d) (see also Fig. 1.8) enables uncracked concrete to sustain the applied shear force. In compliance with this mechanism, the presence of triaxial stress conditions (in localized regions of the compressive zone where the depth is small) delays the development of tensile stresses (caused by the shear force); therefore, the value of the shear force required to cause failure of the compressive zone becomes significantly larger than that expected to cause failure in compliance with the concepts underlying current design methods.

It should also be noted that, in accordance with the experimental data presented in Sect. 1.3, the compressive zone of the element illustrated in Fig. 2.5 is subjected to large axial stresses which, owing to the triaxiality of the stress conditions at the ultimate limit state of the beam, may be over 50 % larger than the uniaxial compressive strength (f_c) of concrete.

It would appear from the above outline of the stress conditions in a typical RC beam, therefore, that, while the magnitude of the stresses that develop within cracked concrete cannot exceed a value of the order of the tensile strength of concrete (i.e. a value of approximately 5–10 % of the strength of concrete in uniaxial compression), the magnitude of the stresses that develop within uncracked concrete can be of the order of the uniaxial compressive strength of concrete or even larger than it by a factor which, in localized regions, may be as large as nearly 2.

2.2.4 Contribution of Uncracked and Cracked Concrete to the Beam's Load-Carrying Capacity

As discussed in the preceding section, uncracked concrete sustains not only the *total* axial compressive force that develops within the beam on account of bending, but also the *total* shear force, the largest portion of which current codes assume to be resisted by cracked concrete through 'aggregate interlock' and 'dowel action'. As a result, the contribution of uncracked concrete essentially represents the *total* contribution of concrete to the load-carrying capacity of the beam.

In contrast, cracked concrete, through the formation of 'plain-concrete cantilevers' between consecutive flexural and/or inclined cracks, provides a mechanism which allows it to make a significant contribution to the transfer of the external load, through the uncracked portion of the beam, from its points of application to the supports. As described in Sect. 2.2.2, this mechanism involves the development of bending moments at the fixed ends of the cantilevers (interface between uncracked and cracked concrete) which balances the actions arising from the shear forces acting at beam cross-sections, including flexural or inclined cracks (as indicated in Fig. 2.4a). As described also in Sect. 2.2.2, the development of the bending moments is attributable to the forces ΔT (see Fig. 2.3) which are transferred from steel to concrete (in the free-end region of the cantilever) through bond.

2.2.5 Causes of Failure

Figure 2.6 shows a schematic representation of the uncracked portion of the beam as a free body under the action of the external load, applied at its top face, and the action of the internal forces developing along the cut which separates the uncracked portion from the remainder of the beam. The figure also provides an indication of the locations where tensile stresses are likely to develop within the uncracked portion.

As discussed in the preceding sections, the uncracked portion of the beam, through which the applied load is transferred to the supports, encloses the path of the compressive stress resultant which develops within the compressive zone due to the bending of the beam. As discussed in Sect. 2.2.2, this transfer requires, on the one hand, the contribution of the cracked portion of the beam through 'cantilever bending' (the latter causes the internal actions which develop at the interface between the uncracked and cracked regions (see Fig. 2.5b and c) and, on the other hand, the change in the path direction (see Fig. 2.4b which occurs at the locations where the middle horizontal narrow strip joins the end blocks of the uncracked portion of the beam (see Fig. 2.6).

From the schematic representation of the distribution of the compressive stresses (σ_c) within the end region of the beam shown in Fig. 2.6, it becomes apparent that only a diagonal strip of this region, which forms essentially an

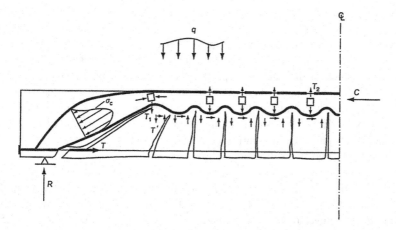

Fig. 2.6 Crack-free portion of simply-supported beam under the action of external load and internal actions at the interface between cracked and crack-free portions of the beam

extension of the compressive zone, is utilized for the transfer of the applied load to the support. With regard to the remaining portion of this end region, its lower part provides anchorage space for the longitudinal reinforcement, while the upper part remains essentially 'structurally' inert, in the sense that it does not make any significant contribution to the transfer of the applied load to the supports, in spite of the development of parasitic stresses of small magnitude and varying orientation.

In addition to its contribution to the transfer of the applied load to the supports through bond-induced cantilever action, the presence of the cracked portion of the beam effects the interaction between uncracked concrete and the longitudinal reinforcement, while, at the same time, it maintains the relative position of the above two components of the beam essentially unchanged throughout the loading history of the beam.

Having established in the preceding section that the uncracked portion of the beam is the sole concrete contributor to the load-carrying capacity of the member (the latter being also dependent on the strength of the longitudinal reinforcement), it is essential to identify the causes of beam failure. If it is assumed that the beam is designed so that it does not suffer any loss of its load-carrying capacity as a result of failure of the longitudinal steel bars, then the causes of failure should be sought in the portion of the beam which comprises uncracked concrete, since cracked concrete could be viewed as concrete already failed.

On the basis of the experimental data presented in the preceding chapter, concrete always fails in tension. As a result, the search for the causes of failure of the portion of the beam comprising uncracked concrete only must be focused on the identification of regions of this portion where tensile forces are likely to develop. Such regions may be the following.

(a) *Regions of change in the direction of the path of the compressive stress result-
    ant.* A tensile stress resultant (T_1 in Fig. 2.6) may develop at the location
    where the path changes direction as a response to the action of the vertical
    component of the inclined compressive stress resultant, developing within
    the end block of the uncracked portion of the beam, which tends to separate
    the upper part of the compressive zone from the remainder of the beam by
    splitting near-horizontally this zone in the region of the change in the path
    direction. (The change in direction of the stress trajectory necessitates, for
    equilibrium purposes, a (nearly-vertical) orthogonal force bisecting the angle
    between the two stress directions.)

(b) *Interface between uncracked and cracked concrete.* As indicated in Fig. 2.6,
    tensile actions (in the sense that they pull the 'cracked' regions away from the
    'uncracked' ones) develop, as described in Sect. 2.2.3, at the above interface
    due to 'cantilever bending' in the cracked region of the beam (see Fig. 2.5b.
    Since, as deduced from the expression $T'z' = Va$ in Fig. 2.5, T' is propor-
    tional to V, an indication of the variation of the magnitude of T' within the
    beam span may be obtained from the shear force. From the latter's diagram,
    it can be seen that the most likely tensile action to cause failure (i.e. σ' to
    exceed f_t) is that (T' in Fig. 2.6) which develops in the region of the tensile
    action T_1 where the compressive-force path (to which the uncracked por-
    tion of the beam forms an envelope) changes direction. Failure in this region
    may occur not only because the tensile action in this region attains the larg-
    est value (as indicated by the shear-force diagrams outside the region of the
    uncracked beam end (in the latter, load transfer does not occur through can-
    tilever bending) of most types of loading condition considered in practice),
    but also because the inclined crack in this region has the most favourable ori-
    entation for crack extension (as opposed to more central portions of the shear
    span, where the existing (mainly flexural) cracks are near-normal to the cracks
    caused by T_1). (The tangent to the shape of the inclined crack at the crack tip
    coincides essentially with the orientation of the principal compressive stress
    which defines the direction of crack extension.)

(c) *Regions adjacent to those including cross-sections with deep flexural or
    inclined cracks.* Volume dilation of concrete in the compressive zone of
    regions including cross-sections with deep flexural or inclined cracks induces
    transverse tensile actions T_2 in the adjacent regions. (A full description of this
    mechanism for the development of such transverse actions is given in Sect.
    1.3.) Four such possible locations are illustrated generally in Fig. 2.6.

(d) *Regions of applied point loads.* These regions usually include cross-sec-
    tions within the shear span where the applied bending moment is large (see
    Fig. 2.7). At the ultimate limit state of the beam, it is likely for bond failure
    to occur in the tensile zone of such regions (see Fig. 2.8). From the figure,
    it can be seen that the loss of bond results in an extension of the right-hand
    side flexural crack sufficient to cause an increase Δz of the lever arm such

Fig. 2.7 Effect of bond loss
on tensile force sustained by
longitudinal reinforcement
(note that the shape of the
variation of the tensile force
before bond loss occurs is
similar to that of the of the
bending moment diagram)

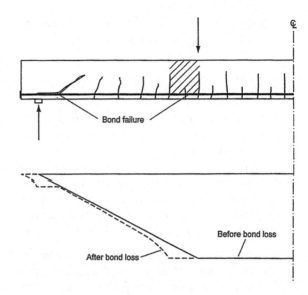

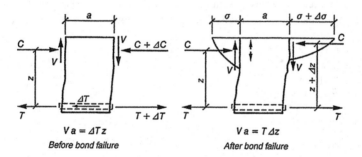

Fig. 2.8 Redistribution of internal actions in the compressive zone due to loss of bond between
concrete and the longitudinal steel bars

that $T\Delta z = Va$ (thus preserving moment equilibrium which would otherwise
be lost because of the elimination of ΔT as a result of the bond destruction).
The extension of the flexural crack reduces the depth of the neutral axis and
this increases locally the intensity of the compressive stress field. In turn, this
increase in the stress intensity should give rise to tensile actions in the manner previously described in item (c) above. (Therefore, one can conclude that
bond failure is likely to occur either near the support (due to the propagation
of the inclined crack towards the support along the interface between concrete
and the longitudinal reinforcement) or at locations of large bending moment
(and non-zero shear) because of the tensile yielding of the bars.)

2.3 Validity of Proposed Structural Functioning of Simply-Supported Beams

The description of the functioning of the simply-supported beam proposed in the preceding sections contrasts with current views with regard to the following points.

(a) Uncracked concrete in compression (through which the applied load is transferred to the supports) and the longitudinal main-steel bars in tension are essentially the sole contributors to the load-carrying capacity of the beam, with cracked concrete contributing mainly to the transfer of applied load to the supports through 'cantilever bending'.

(b) Failure of the beam is caused by the development of tensile stresses within the previously uncracked concrete, which act transversely to the longitudinal compression that develops as a result of the bending of the beam.

The adoption of point (a) above is fully justified as this premise is compatible with the experimental data presented in the preceding chapter. However, point (b) remains to be proved as compatible with experimental data available to date on beam behaviour at the ultimate limit state. Such data have been summarized in Fig. 2.9, which provides a schematic representation of the variation of the load-carrying capacity of a simply-supported RC beam, without stirrups, under

Fig. 2.9 Characteristic types of behaviour of a simply supported beam at its ultimate limit state. Modes of failure (*top*) and relation between bending moment corresponding to load-carrying capacity and shear span for various percentages of longitudinal reinforcement (*bottom*)

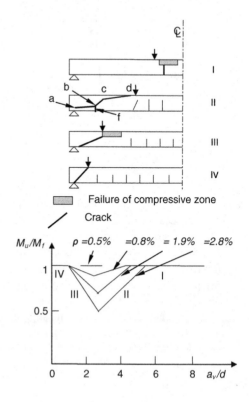

two-point loading with the shear span-to-depth (α_v/d) ratio for various percentages of the longitudinal reinforcement, with the beam's load-carrying capacity being expressed in the form of the bending moment at the mid cross-section. (This ratio M_u/M_f reflects the actual capacity of the beam relative to its full flexural capacity.) From this form of representation of the data (first introduced by Kani [1]), it becomes apparent that the behaviour of the above beam, at its ultimate limit state, may be divided into four types of regimes associated with the value of α_v/d.

Type I behaviour

Type I behaviour corresponds to relatively large values of α_v/d (usually larger than 5) and is characterised by a flexural mode of failure. The causes of such a mode of failure are fully described in Sect. 1.3 of the preceding chapter and they have already been incorporated into the proposed qualitative description of beam behaviour (see item (c) in Sect. 2.2.5).

Type II behaviour

Type II behaviour corresponds to values of α_v/d between approximately 2.5 and 5, and is characterised by a brittle mode of failure which is usually associated with the formation of a deep inclined crack within the shear span of the beam. (Brittle failure may also occur owing to near-horizontal splitting of the compressive zone which occurs independently from any web cracking in the region of the point load, as discussed later in this section.) Immediately after its formation, the inclined crack (which, for values of α_v/d closer to 2.5 rather than 5, is essentially an extension of the flexural crack (marked with *f* in Fig. 2.9) closest to the support) extends near-horizontally (branch *c-d* in Fig. 2.9) within the compressive zone towards the point load in an unstable manner, leading to an immediate and total loss of load-carrying capacity of the beam. (This inclined crack may also extend towards the support along the interface between concrete and the steel bars (branch α-*b* in Fig. 2.9), destroying the bond between the two materials, but such an extension may be prevented from leading to failure of the beam by proper anchoring of the steel bars.)

The causes of such a mode of failure are described by items (a) and (b) in Sect. 2.2.5. These are associated with the development of tensile actions in the region where the path of the compressive force (owing to the bending of the beam) changes direction. Such tensile actions, as discussed in Sect. 2.2.5, may cause splitting of the compressive zone which leads to total loss of the beam load-carrying capacity. Details of the manner in which the above failure process initiates is illustrated in Fig. 2.10 (left) and (right) which show that, between the upper face of the beam and the inclined crack closest to the support, in the region of the crack tip, an isolated (deepest) crack (marked with c in the two figures) forms as soon as the tensile strength of concrete is exhausted. [The extension of this crack was prevented by the instantaneous unloading of the beam as soon as the crack appeared. Maintaining the load constant leads to the failure process described above, which gives the (misguided) impression that failure is caused by the extension of the inclined crack, as usually depicted for type II behaviour (as in Fig. 2.9)]. It becomes apparent from the above, therefore, that in order to prevent this type of failure the location at which the compressive force changes direction must be known α *priori*.

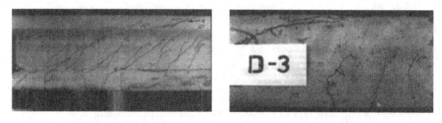

Fig. 2.10 Horizontal cracking which precedes failure of the compressive zone of the beam in Fig. 1.6 under two-point loading (*left*) and a beam similar to beam D1 in Fig. 1.2b (*right*)

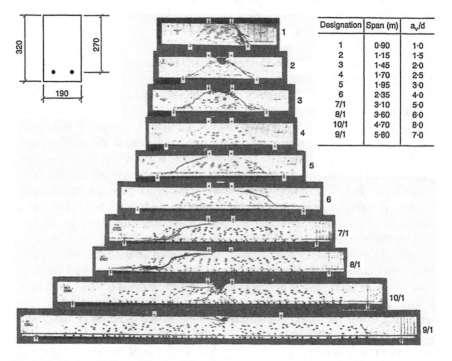

Designation	Span (m)	a_v/d
1	0·90	1·0
2	1·15	1·5
3	1·45	2·0
4	1·70	2·5
5	1·95	3·0
6	2·35	4·0
7/1	3·10	5·0
8/1	3·60	6·0
10/1	4·70	8·0
9/1	5·80	7·0

Fig. 2.11 Modes of failure of beams with various shear spans under two-point loading [2]

Figures 2.11 and 2.12 illustrate the crack patterns of two groups of beams at failure tested under two-point and uniformly-distributed loading respectively [2]. As discussed in Sect. 2.2.2, the change in the direction of the path of the compressive force which develops due to the bending of the beam occurs in the region of the tip of the inclined crack forming closest to the support. For the beams under two-point loading with values of a_v/d between 2.5 and 5 (beams 4–6 in Fig. 2.11), as well as for the beams under uniformly-distributed loading with a normalised (with respect to the beam depth) span (L/d) greater than 8 (beams 13–17 in Fig. 2.12), the

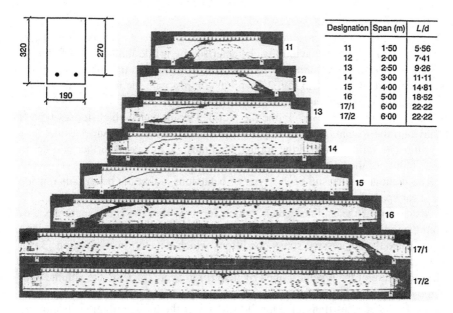

Designation	Span (m)	L/d
11	1·50	5·56
12	2·00	7·41
13	2·50	9·26
14	3·00	11·11
15	4·00	14·81
16	5·00	18·52
17/1	6·00	22·22
17/2	6·00	22·22

Fig. 2.12 Modes of failure of beams with various spans under uniformly-distributed loading [2]

tip of the above crack lies at a distance approximately equal to two and half times the beam depth (2.5d) from the support. It should be expected, therefore, that the provision of sufficient reinforcement at a distance of 2.5d from the supports should prevent beam failure associated with the causes of failure described by items (α) and (b) in Sect. 2.2.5. In fact, placing such reinforcement in beam D1 shown in Fig. 1.2b allowed the beam to develop its flexural capacity in contrast to the predictions of current methods used for the design of reinforced-concrete structures.

As discussed earlier, beams characterised by type II behaviour may also fail as a result of horizontal splitting of the compressive zone which occurs independently from any web crack. Such splitting may be due to the development of tensile stresses within the compressive zone associated with the loss of bond between concrete and flexural steel as described by item (d) in Sect. 2.2.5. Loss of bond may have been the cause of failure of beams 7 and 8 in Fig. 2.11 whose mode of failure is also characterised by the presence of an inclined crack which formed closer to the load point rather than the support. Although bond failure appears to have occurred in the region between the deep inclined crack and the crack adjacent to it (as one moves away from the support), it cannot be deduced from the modes of failure indicated in the figure that the failure process was that predicted by item (d) in Sect. 2.2.5. Additional information regarding this type of failure will be provided in Chap. 3 as part of the verification study of the new design methodology proposed in that chapter.

Type III behaviour

Type III behaviour corresponds to values of a_v/d between approximately 1 and 2.5 and, as for type II behaviour, is characterised by brittle failure. Such failure is associated with the development of an inclined crack within the shear span of the beam which, as indicated in Fig. 2.9, and in contrast with the inclined crack characterising the type II behaviour, forms independently from the pre-existing flexural or inclined crack. Moreover, unlike the inclined crack which characterises type II behaviour, the formation of an inclined crack for type III behaviour does not lead to immediate failure; instead the applied load must be increased further in order to cause failure of the beam.

The main characteristic of this type of behaviour is that the beam fails outside the shear span. As indicated by the mode of failure of beams 2 and 3 in Fig. 2.11, the extension of the inclined crack, which forms within the shear span, deviates from the region of the applied load, where the strength of concrete is higher owing to the triaxial compressive stress conditions which develop in this region [3, 4], and penetrates deeply into the compressive zone of the 'flexure' span of the beam causing failure of the type described by item (c) in Sect. 2.2.5, i.e. the volume dilation of concrete in the compressive zone of the cross-section through the tip of the inclined crack causes transverse tensile stresses in the adjacent regions leading to splitting of the compressive zone and failure of the member, before yielding of the flexural reinforcement.

The above explanation of the causes of failure is compatible with the experimental data obtained from the tests on the beams of type D shown in Fig. 1.2a. From this figure, it appears that placing links only within the 'flexure' span of beams with $a_v/d = 1.6$ delays the formation of horizontal cracks in the region of the point load sufficiently for the beam to exhaust its flexural capacity first. It is also of practical interest, as will be seen in Chap. 3, to note that for the case of point loading the change in the direction of the path of the compressive stress resultant occurs in the cross-section through the point load (see beams 2 and 3 in Fig. 2.11), whereas for the case of uniformly-distributed loading this change in path direction occurs at a distance from the support approximately equal to a quarter of the beam span (see beams 11 and 12 in Fig. 2.12).

Type IV behaviour

Type IV behaviour corresponds to values of a_v/d smaller than 1 and is characterised by two possible modes of failure [5]: (a) a ductile mode of failure, for the case of failure within the middle narrow strip of the uncracked portion of the beam; and (b) a brittle mode of failure, for the case of failure of the end blocks of the uncracked portion of the beam in the region of the support. As will be seen in Chap. 3, the mode of failure is generally dictated by the size of the beam width, the larger sizes being more likely to lead to a ductile, rather than a brittle, type of failure. It should be noted, however, that in both cases there is no significant, if any, difference in load-carrying capacity. The mechanism of failure described by item (c) in Sect. 2.2.5 provides a satisfactory description of the causes of failure which characterises the present type of behaviour.

2.4 Conclusions

The present chapter summarises the experimental data presented in Chap. 1 in a manner that reveals the fundamental characteristics which underlie the behaviour of a simply-supported RC beam, without stirrups, at its ultimate limit state under static monotonic transverse loading. The main conclusions drawn from this summary are as follows:

1. The beam comprises the following:
   (a) an uncracked portion consisting of the two end-regions of the beam which, extending to the (usually) inclined crack forming closest to the support and the cross-section through the tip of this crack, are connected by a narrow strip of varying depth forming between the upper face and the tips of flexural and inclined cracks which initiate at the bottom face and extend towards the upper face of the beam,
   (b) a cracked portion consisting of 'plain-concrete cantilevers' which, forming between successive flexural and inclined cracks, are fixed at the narrow zone of the uncracked portion,
   (c) the longitudinal reinforcement, penetrating the beam throughout its span at a relatively short distance from the tensile face, fully bonded to concrete at least in the region of the support where it is properly anchored.

2. The uncracked portion encloses the path of the compressive stress resultant (due to the bending of the beam), with a near-horizontal orientation within the middle narrow strip of the uncracked portion, changing in the region of the tip of the inclined crack closest to the support and becoming diagonal within the end regions. The location of the change in the path direction appears to depend on parameters such as, for example, the shear span-to-depth-ratio for the case of point loading, and the span-to-depth ratio for the case of uniformly-distributed loading.

3. The uncracked portion of the beam is not only the sole concrete contributor to the load-carrying capacity of the member but also transfers the applied load to the supports; the cracked portion of the beam contributes to this transfer through cantilever bending.

4. Failure appears to be associated with the development of transverse tensile stresses within the uncracked portion of the beam. The causes for the development of such stresses vary and appear to depend on the value of the parameters referred to in conclusion 2 above.

References

1. Kani GNJ (1964) The riddle of shear failure and its solution. ACI J Proc 61(28):441–467
2. Leonhardt F, Walther R The Stuttgart shear tests, 1961, contributions to the treatment of the problems of shear in reinforced concrete construction. (A translation (made by C.V. Amerongen) of the articles that appeared in *Beton-undStahlbetonball*, Vol. 56, No. 12, 1961, and Vol. 57, Nos. 2-3, 7-8, 1962.) Translation No. 111, C&CA, London, 1964

3. Kotsovos MD, Newman IB (1981) Effect of boundary conditions upon the behaviour of concrete under concentrations of load. Mag Concr Res 33(116):161–170
4. Kotsovos MD (1981) An analytical investigation of the behaviour of concrete under concentrations of load. Mater Struct RILEM 14(83):341–348
5. Kotsovos MD (1988) Design of reinforced concrete deep beams. Struct Eng 66(2):18–32

Chapter 3
Modelling of Simply-Supported Beams

3.1 Introduction

In this chapter, the qualitative description of the beam behaviour presented in Chap. 2 is transformed into a physical model with behavioural characteristics (such as, for example, crack pattern, internal actions, transfer mechanism of external load to the supports, failure mechanism, etc.) similar to those of a real simply-supported beam at its ultimate limit state. The physical model's behavioural characteristics form the basis for the development of failure criteria which are shown to be capable of providing realistic predictions of a beam's load-carrying capacity for all types of behaviour discussed in the preceding chapter.

3.2 Physical Model

Figure 3.1a depicts the physical model of a simply-supported beam, the qualitative characteristic features of which were described in detail in the preceding chapter. Figure 3.1b provides a schematic representation of the effect that the presence of external axial load has on this physical model. As indicated in Fig. 3.1a, the beam, in all cases, is modelled as a 'comb-like' structure with 'teeth' fixed on to the horizontal element of a frame with inclined legs. The 'frame' and the 'teeth' also interact through a horizontal 'tie' which is fully bonded to the 'teeth', near their bottom face, and anchored at the bottom ends of the 'frame' legs. A comparison between the proposed model and the beam of Fig. 2.1 indicates the following.

(a) The 'frame' provides a simplified representation of the uncracked region of the beam which encloses the path of the compressive-stress resultant that develops due to bending.
(b) The 'tie' represents the flexural reinforcement.

M. D. Kotsovos, *Compressive Force-Path Method*, Engineering Materials,   41
DOI: 10.1007/978-3-319-00488-4_3, © Springer International Publishing Switzerland 2014

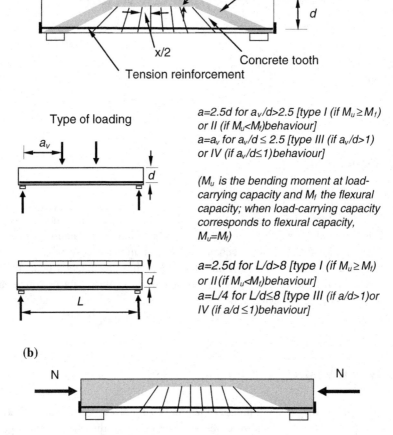

Fig. 3.1 a Physical model of simply-supported RC beam under transverse loading and **b** effect of axial load on physical model in (**a**)

(c) The 'teeth' of the 'comb-like' model represent the plain-concrete cantilevers which form between successive flexural or inclined cracks within the tensile cracked zone of the beam.

As concluded in the preceding chapter, the load-carrying capacity of the beam is provided by the combined action of the uncracked concrete and the flexural reinforcement, i.e. the 'frame' and the 'tie' of the proposed model, with uncracked concrete, i.e. the 'frame', also transferring the applied load to the supports, while cracked concrete in the tensile zone, i.e. the 'teeth' of the 'comb', provides the (bond-based) mechanism through which the transfer loop is completed.

3.3 Failure Criteria

In order to implement in practical design the physical model presented in the preceding section, it is essential to complement it with failure criteria capable of predicting both load-carrying capacity and mode of failure. Such failure criteria must be compatible with experimental information such as that summarised, in a pictorial form, in Fig. 2.9. The figure includes a graphical description of the beam load-carrying capacity, together with schematic representations of the modes of failure characterising four distinct types of behaviour indicated in the figure.

Although, as discussed in Sect. 2.3, a common feature of all types of behaviour is that failure appears to be associated with the development of transverse tensile stresses within the uncracked portion of the beam, i.e. the 'frame' of the proposed model, the causes for the development of these stresses and the locations of failure differ for each type of behaviour. In view of this, the failure criteria proposed in what follows have been developed so as to reflect the causes of failure relevant to each of the four distinct modes of failure indicated in Fig. 2.9.

3.3.1 Type I Behaviour

As discussed in Sect. 2.3, type I behaviour is characterised by a flexural mode of failure which is preceded by longitudinal splitting of concrete in the compressive zone of the beam, i.e. the horizontal member of 'frame'; splitting occurs when concrete strength is exhausted under the action of transverse tensile stresses induced by volume dilation of concrete in adjacent regions which include primary flexural cracks. (A full description of this mechanism for the development of such transverse actions is given in Sect. 2.3.) The methods currently used for calculating flexural capacity do not allow for such splitting and this is considered to be the reason that they are invariably found to underestimate (often by a significant margin) the flexural capacity of linear reinforced concrete (RC) elements, such as beams and columns [1]. This problem is widely recognized in RC capacity design and current code provisions aiming to safeguard against shear types of failure recommend the use of an "over-strength" factor which leads to an increase in flexural capacity up to 40 % in certain cases [2]. The causes of the underestimate are commonly attributed to discrepancies between the true and assumed material properties, and, in particular, to the strain hardening of the steel reinforcement, which is not usually allowed for in the calculation [3].

Allowing for strain hardening results in an increase of the tensile force sustained by the flexural reinforcement, which, for purposes of internal force equilibrium, must be balanced by an increase of the force sustained by concrete in the compressive zone. Since the calculation of flexural capacity is based on the assumption of a uniaxial stress field in the compressive zone, the increase of the force sustained by concrete can only occur through an increase of the depth of the compressive zone.

Fig. 3.2 Schematic
representation of the stress
conditions in the compressive
zone of the portion of an RC
beam in pure bending

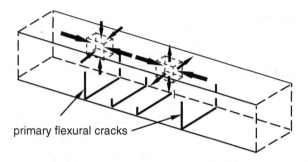

primary flexural cracks

The latter increase inevitably leads to a reduction of the internal force lever arm
and this sets an upper limit to the resulting increase in flexural capacity. As a result,
even when allowing for strain hardening, current methods of calculation may still
underestimate flexural capacity.

(a) *Assessment of stress conditions in compressive zone*

However, as the experimental results presented in Sect. 1.3 indicate, adopting uniaxial
stress-stain characteristics is incompatible with the measured deformational response of
concrete in the compressive zone of an RC beam (without compression reinforcement
and stirrups) at its ultimate limit state in flexure. Moreover, the stress conditions within
this zone are triaxial rather than uniaxial; in fact, these stress conditions are wholly
compressive in the regions of cross-sections including primary flexural cracks, whereas
between such regions longitudinal compression is combined with transverse (both in
the vertical and in the horizontal directions) tension as indicated in Fig. 3.2 [4].

The relationship between axial (σ_a) and transverse confining (σ_c) stresses may
be expressed by [1]

$$\sigma_a = f_c + 5\sigma_c \tag{3.1}$$

with f_c being the uniaxial cylinder compressive strength of concrete.

This expression has been found to provide a close fit to experimental data
such as those shown in Fig. 1.11 for values of σ_c equal to up to about $0.5f_c$ [1].
Moreover, it has recently been shown that, just before flexural failure occurs, the
transverse confining stresses (σ_c) developing in the compressive zone obtain values
which, numerically, can be as large as the tensile strength of concrete [1]; the latter
may be obtained from the following expressions [5]:

$$f_t = f_{to} \, (f_{ck}/f_{cko})^{2/3} \text{ for normal-strength concrete} \tag{3.2a}$$

where

$f_{ck} = f_c - 8$ (with f_c being the uniaxial compressive strength of concrete),
$f_{cko} = 10, f_{to} = 1.4$ (mean value), 0.95 (minimum value) and 1.85 (maximum)

$$f_t = f_{to} \ln \left(1 + f_c/f_{co}\right) \text{ for high-strength concrete } (f_{ck} \geq 60 \, \text{MPa}) \tag{3.2b}$$

where $f_{co} = 10, f_{to} = 2.12$ (mean value), with all values of strength expressed in MPa.

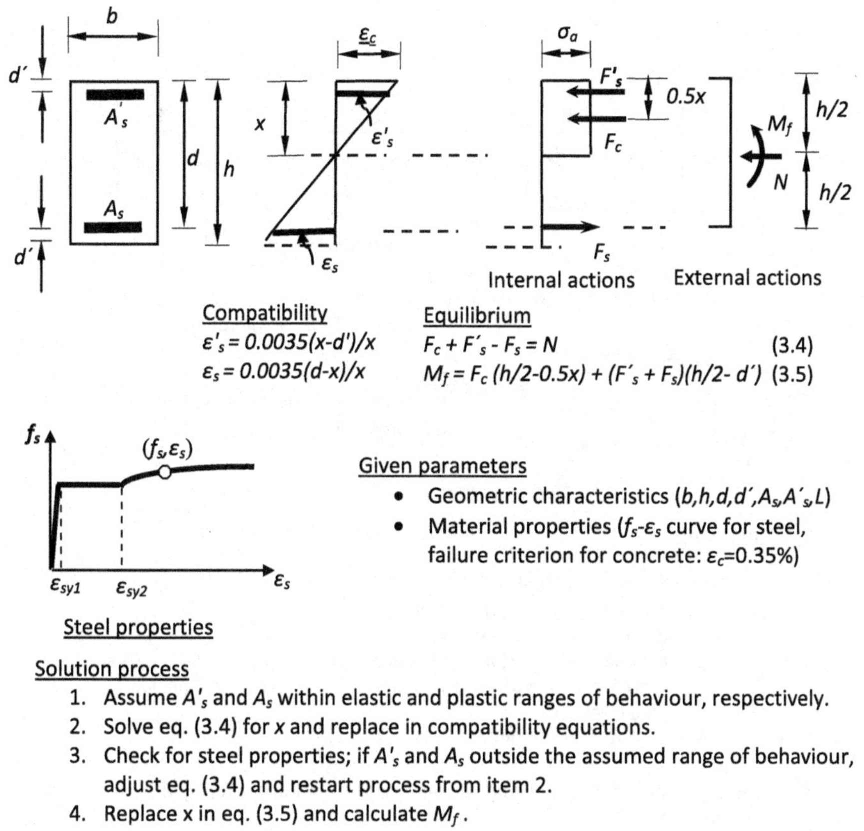
b
d′
A′s
As
d
h
x
εc
ε′s
εs
σa
F′s
0.5x
Fc
Fs
Mf
N
h/2
h/2
Internal actions
External actions
Compatibility
ε′s = 0.0035(x-d′)/x
εs = 0.0035(d-x)/x
Equilibrium
Fc + F′s - Fs = N (3.4)
Mf = Fc (h/2-0.5x) + (F′s + Fs)(h/2- d′) (3.5)
fs
(fs,εs)
εsy1
εsy2
εs
Steel properties
Given parameters
Geometric characteristics (b,h,d,d′,As,A′s,L)
Material properties (fs-εs curve for steel, failure criterion for concrete: εc=0.35%)
Solution process
1. Assume A′s and As within elastic and plastic ranges of behaviour, respectively.
2. Solve eq. (3.4) for x and replace in compatibility equations.
3. Check for steel properties; if A′s and As outside the assumed range of behaviour, adjust eq. (3.4) and restart process from item 2.
4. Replace x in eq. (3.5) and calculate Mf.

Fig. 3.3 Proposed method for calculating flexural capacity

(b) *Calculation of flexural capacity*

The method proposed for the calculation of flexural capacity is a further simplified version of the method proposed in Ref. [1]; it is described in Fig. 3.3 which shows that it is essentially the method adopted by the current European code (EC2 [3]) modified so as to allow for the development of triaxial stress conditions in the compressive zone. This modification involves the replacement of the stress intensity $0.85f_c$ of the code specified simplified stress block (describing the state of stress of concrete in the compressive zone) with σ_α, as obtained from expression (3.1), with the stress block extending throughout the depth x of the compressive zone, rather than the code specified depth of $0.8x$. Since, as discussed in the preceding section, just before flexural failure σ_c becomes numerically equal to f_t, σ_α can be calculated through the use of expression (3.1) by replacing σ_c with the absolute value of f_t, the later resulting from expression (3.2a) or expression (3.2b), i.e.

$$\sigma_a = f_c + 5\,|f_t| \tag{3.3}$$

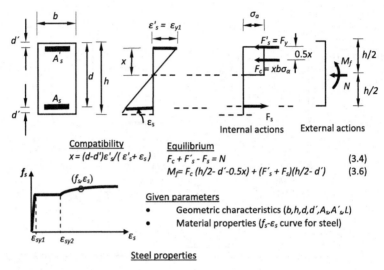

Compatibility
$x = (d-d')\varepsilon'_s /(\varepsilon'_s + \varepsilon_s)$

Equilibrium
$$F_c + F'_s - F_s = N \qquad (3.4)$$
$$M_f = F_c \,(h/2 - d' - 0.5x) + (F'_s + F_s)(h/2 - d') \qquad (3.6)$$

Given parameters
- Geometric characteristics (b,h,d,d',A_s,A'_s,L)
- Material properties $(f_s\text{-}\varepsilon_s$ curve for steel$)$

Steel properties

Solution process
1.   Set $\varepsilon_s = \varepsilon_{y2}$
2.   Calculate x from compatibility equation
3.   Check whether equation (3.4) is satisfied.
   -   If it does, continue to step 4;
   -   If it does not, solve (3.4) for ε_s and obtain corresponding f_s from f_s - ε_s
       curve for steel; return to step 2 and repeat process until equation
       (3.4) is satisfied.
4.   Calculate M_f from expression (3.6)

Fig. 3.4 Proposed method for calculating flexural capacity allowing for spalling of cover to compression reinforcement

In fact, expressions (3.2a) and (3.2b) may be used as the means to control the margin of safety against a non-flexural (brittle) type of failure by selecting a suitable value for parameter f_{to}. For example, in situations where safeguarding ductile behaviour is of paramount importance (design for earthquake resistant structures), the maximum value of f_{to} may be adopted, as opposed to the mean value that may be more appropriate in ordinary design.

It should be noted that for flexural failure to be ductile, A_s should yield before the strain of concrete at the extreme compressive fibre reaches its limiting value $\varepsilon_c = 0.35 \%$ (see Fig. 3.3). It should also be noted that failure of concrete in the compressive zone is preceded by longitudinal cracking and spalling (see Fig. 3.6b) which reduces significantly the size of this zone, and, in the absence of compression reinforcement, leads to a considerable, if not complete, loss of load-carrying capacity. On the other hand, the presence of compression reinforcement may delay this failure process until yielding and subsequent buckling of the compression steel bars (see Fig. 3.6c). In order to allow for this type of failure, the proposed method of calculation of the flexural capacity is complemented with the procedure described in Fig. 3.4. The specimen's load-carrying capacity corresponds to the larger of the values resulting from the procedures described in Figs. 3.3 and 3.4.

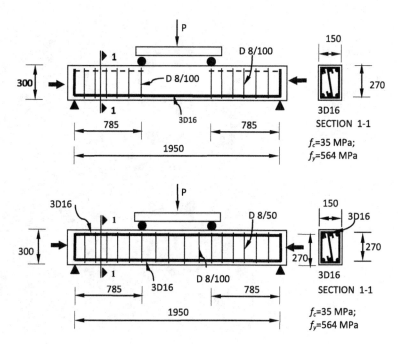

Fig. 3.5 Design details of two typical beam-column specimens

(c) *Verification of proposed method*

The investigation of the validity of the method proposed for calculating flexural capacity is essentially based on a comparative study of predicted values with their experimentally established counterparts obtained from tests on simply-supported beams under two-point loading. The design details of two typical such beams are shown in Fig. 3.5, with the crack pattern at various stages of the load applied on the beam in Fig. 3.5b being shown in Fig. 3.6. The results of this comparative study are summarised in Table 3.1, whereas full details are provided elsewhere [1].

Table 3.1 shows the experimentally established mean values of load-carrying capacity together with their calculated counterparts before (column 6) and after (column 8) the implementation of the proposed modifications. From the table, it can be seen that the proposed method yields predictions whose deviation from the experimentally-established values is on average of the order of 2 %. This is a significant improvement over the predictions of the code adopted methods, since the difference between the latter predictions and those of the proposed method can be as large as 24 % in certain cases (see column 9 of the table).

3.3.2 Type II Behaviour

The brittle modes of failure associated with type II behaviour (encompassing, approximately, the range of a_v/d between 2.5 and 5) is caused by tensile stresses developing either in the region of change of the CFP direction (location 1 within

Fig. 3.6 Crack patterns
of a typical specimen at
characteristic stages of its
behaviour: **a** formation
of longitudinal cracks in
compressive zone; **b** spalling
of concrete covering the
compression bars; and **c**
buckling of the compression
bars

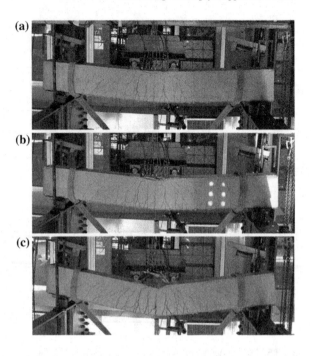

shear span a_{v1}, in Fig. 3.7, assuming $a_{v1} \geq 2.5d$) or in the region of the cross-section
at the left-hand side of the point load, where the maximum bending moment combines with the shear force (location 2 within shear span a_{v1}, in Fig. 3.7). By invoking St. Venant's principle, the effect of the transverse stress resultant at location 1
spreads to a distance equal to the cross-section depth d, on either side of location 1,
where the CFP changes direction. On the other hand, as discussed in Sect. 2.3, transverse tensile stresses within the compressive zone of the cross-section where the
maximum bending moment combines with the shear force (location 2 in Fig. 3.7)
may develop due to the loss of bond between the longitudinal reinforcement and the
surrounding concrete in the manner indicated in Fig. 3.8.

Figure 3.8 indicates a portion of the structural element between two cross-sections
defined by consecutive cracks which, on the basis of experimental data such as
those shown in Figs. 2.11 and 2.12, have a spacing $a \approx x/2$ (where x is the depth
of the compressive zone), together with the internal forces which develop at these
cross-sections before and after the loss of bond τ necessary to develop due to the
increase in tensile force ΔF_S. Setting the flexural moment $M = F_s z$ and observing
that the shear force $V = dM/dx = (dF_s/dx) z + F_s (dz/dx)$, it can be seen that the
two products on the right-hand side of the equation correspond to beam (bond)
and arch (no bond) action, respectively. From the figure, it can be seen that the
loss of bond may lead to an extension of the right-hand side crack, and hence a
reduction of the compressive zone depth (x), which is essential for the rotational
equilibrium of this portion as indicated by the relation

Table 3.1 Experimental values of load-carrying capacity of statically determinate RC members together with their counterparts predicted by the code and the proposed herein methods

Spec	f_c (MPa)	f_y (MPa)	P_{exp} (kN)	$P_{f,code}$ (kN)	$P_{f,code}/P_{exp}$	$P_{f,proposed}$ (kN)	$P_{f,proposed}/P_{exp}$	$10^2 \cdot [(8)-(6)]/(6)$
(1)	(2)	(3)	(4)	(5)	(6)	(7)	(8)	(9)
Reference 1								
1-0a	35	564	208	199.6	0.960	213.8	1.027	6.98
1-0b			202		0.988		1.058	7.09
1-0c			212		0.941		1.008	7.12
1-150			241.8	221.3	0.915	249.3	1.003	9.62
1-250a			256	228.6	0.893	268.7	1.050	17.56
1-250b			258.8		0.883		1.038	17.55
2-0a			250	211.2	0.845	255[*]	1.020	24.5
2-0b			242		0.873		1.054	20.7
2-150			258.2	253.5	0.982	259. 7	1.006	2.44
2-250a			283	278.6	0.984	287.1	1.005	2.13
2-250b			286.5		0.972		1.002	3.09
Average					**0,931**		**1,024**	**10.79**
Stdev					**0,051**		**0,022**	**7,90**
Reference 29								
CFP-10-M	35	621	150	124	0.826	134	0.893	8.1
EC-10-M			148		0.838		0.905	8
CFP-10-C			141		0.879		0.950	8.1
EC-10-C			136		0.946		0.985	4.1
CFP-12-M		554	182	158	0.868	167[*]	0.918	5.8
EC-12-M			188		0.840		0.888	5.7
CFP-12-C			166		0.952		1.001	5.1
EC-12-C			171		0.924		0.977	5.7
Average					**0,884**		0,940	**6,3**
Stdev					**0,050**		0,045	**1,54**
Reference 30								
CFP-0	60	540	102.5	95.7	0.934	105[*]	1.024	9.6
EC-0			100		0.957		1.050	9.7
CFP-400			203.4	188.3	0.925	194.4	0.956	3.4
EC-400			201.4		0.935		0.965	3.2
CFP-500			224,4	207.1	0.923	216.3	0.964	4.5
CFP-675			264.7	232.5	0.878	250.4	0.946	7.7
Average					**0,925**		0,984	**6,35**
Stdev					**0,026**		0,042	**3,02**

[*]Values resulting from procedure in Fig. 3.4; in all other cases, values resulted from procedure in Fig. 3.3

Fig. 3.7 Physical model of a
simply-supported RC beam at
its ultimate limit state under
point loading

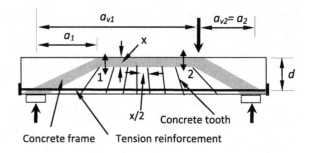

Fig. 3.8 Redistribution
of internal actions in the
compressive zone due to loss
of bond between concrete and
longitudinal reinforcement

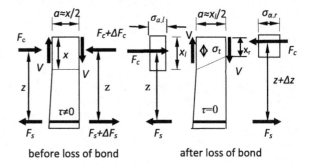

$$F_c \, (x_l - x_r) \, /2 = V(x_l/2) \tag{3.7}$$

where x_l and x_r are the values of x, after loss of bond, at the left- and right-hand
sides, respectively, of the beam element shown in Fig. 3.8.

The reduction of the compressive zone depth increases the intensity of the compressive-stress field as compared to its value at the left-hand side of the portion,
thus leading to dilation of the volume of concrete which causes the development
of transverse tensile stresses (σ_t in Fig. 3.8) in the adjacent regions.

(a) *Calculation of shear capacity of region of location 1*

As proposed in Ref. [6], the transverse tensile stress distribution in the region of
change in the direction of the compressive-stress resultant developing due to bending may be schematically represented as indicated in Fig. 3.9. From the figure,
it can be seen that, just before failure, the stress distribution is characterised, on
the one hand, by a peak value which is equal to the tensile strength (f_t) of concrete (at the point where the horizontal and inclined directions of the compressive-
stress resultant intersect) and, on the other hand, by a rapid reduction of the tensile
stresses with the distance from the peak-stress point. As discussed in the preceding
section, these tensile stresses (by invoking the principle of St. Venant) may be considered to become negligible beyond a distance equal to the effective depth d from
the location of their peak value.

As the exact shape of the above stress distribution is difficult, if not impossible,
to assess, the resulting stress resultant has been assumed to be equivalent to that
of a uniform stress distribution with intensity $0.25f_t$ spreading over the same area

Fig. 3.9 Schematic
representation of transverse
tensile stress distribution
in the region of change in
direction of the compressive-
stress resultant due to
bending and equivalent stress
block

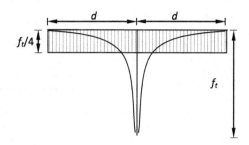

$b(2d)$ (where b is the width of the beam) [6]; then the transverse tensile force that may be sustained by concrete in this region is equal to

$$T_{II,1} = 0.5bdf_t \qquad (3.8)$$

It should be noted that the value of stress intensity ($0.25f_t$) adopted above is typical of the value obtained by integrating a stress distribution such as that in Fig. 3.9 over the area $2bd$ and dividing the result by this area. Since, as indicated in Fig. 2.4b and discussed elsewhere [6], $T_{II,1}$ may be viewed as the vertical component of the inclined compressive-force transmitted to the support, it represents the value of the shear force $V_{II,1}$ that can be sustained within the shear span just before horizontal splitting of concrete in the region where the compressive-force path changes direction; in other words Eq. (3.8) may take the form

$$V_{II,1} = 0.5bdf_t \qquad (3.9)$$

The tensile strength of concrete f_t in expression (3.9) can be calculated from expressions (3.2a) or (3.2b).

When the beam is also subjected to axial compression (i.e. $N \neq 0$), then, the slope of the inclined compression at location 1 may be considered to decrease as indicated in Fig. 3.10 [7]. Then, it has been subjected [7] that

$$T_{II,1,0}/T_{II,1,N} = tan\varphi_o/tan\varphi_N \qquad (3.10)$$

where φ_o, φ_N are the values of the slope of the axial compression and $T_{II,1,0}$, $T_{II,1,N}$ the values of transverse tension developing at location 1 for the cases $N=0$ and $N\neq0$, respectively.

With $T_{II,1,0}$ being equivalent to the reaction (shear force next to the support), expression (3.10) indicates that $T_{II,1,N}$ is smaller than $T_{II,1,0}$ by a factor of $tan\varphi_N/tan\varphi_o$. In view of this, it has been suggested that the presence of an axial force reduces transverse tension at location 1, and, hence, a larger transverse load is required to increase it to its limiting value assessed from expression (3.8) [7]. Since, for $N = 0$, the latter value has been found to be equal to the experimentally established shear capacity ($V_{N=o}$) [6], it has been proposed that, for $N \neq 0$, the value of shear capacity ($V_{N\neq o}$) corresponding to the value of $T_{II,1}$ resulting from expression (3.8) is that resulting from expression (3.9) magnified by a factor $k = tan\varphi_o/tan\varphi_N$ [7] i.e.

Fig. 3.10 Part of the shear span of an RC beam indicating the path of the compressive-force due to bending for the cases $N = 0$ and $N \neq 0$

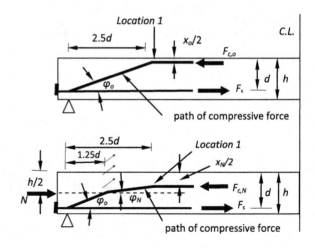

$$V_{N \neq o} = kV_{N=o} = kV_{II,1} = k\,T_{II,1} = k0.5b\,d\,f_t \qquad (3.11)$$

where $tan\varphi_o = (2d - x_o)/5d$ and $tan\varphi_N = tan\varphi_o\,(h - x_N)/(h - x_o)$, thus $k = (h - x_o)/(h - x_N)$, with d, h, x_N, x_o being defined in Fig. 3.10.

(b) Calculation of shear capacity of region of location 2

As discussed earlier, failure may also occur due to the development of transverse tensile stresses in the compressive zone of the region of the shear span, indicated in Fig. 3.8, subjected to the largest acting bending moment. The development of transverse tensile stresses occurs when the acting bending moment, as it approaches the cross-section's flexural capacity, causes yielding of the tension reinforcement which renders loss of bond inevitable.

The loss of bond disrupts the rotational equilibrium of the element in Fig. 3.8. However, equilibrium can be restored through an increase of the bending moment acting at the right-hand side of the element due to an extension of the flexural crack deeper into the compressive zone; thus the compressive zone depth (x_r) decreases, while the lever arm (z) of the internal actions increases by Δz. The smaller x_r, combined with an increase of the stress intensity ($\sigma_{\alpha,r}$), maintains the compressive-stress resultant (F_c) constant throughout the length of the element, and thus the force equilibrium condition in the longitudinal direction is satisfied, since $F_c = F_s$ on both sides of the element.

Failure occurs when the flexural capacity at the right-hand side of the element is exhausted when, as discussed earlier, σ_t tends to exceed f_t at the left-hand side of the element (see Fig. 3.8), where $\sigma_{a,l}$ is considered to remain unchanged since yielding of the tension reinforcement; for beams designed to exhibit ductile flexural behaviour, yielding usually occurs when $\sigma_{a,l} \approx f_c$. Since the value of F_c is constant throughout the length of the element, $F_c = f_c b x_l = \sigma_a b x_r$, the latter condition leading to

$$x_r = x_l\,(f_c/\sigma_a) \qquad (3.12)$$

Then, the shear force ($V_{II,2}$) that can be sustained by the beam just before horizontal splitting of the compressive zone (when f_t is attained) may be obtained by solving for V expression (3.7) in which x_r/x_l is expressed as a function of f_c by replacing in expression (3.12) $\sigma_{\alpha,r}$ as described in expression (3.3), i.e.

$$V_{II,2} = F_c \left[1 - 1/(1 + 5\,|f_t|\,/f_c)\right] \tag{3.13}$$

As for expression (3.9), the tensile strength of concrete f_t in expression (3.13) can be calculated from one of expressions (3.2a) or (3.2b).

In contrast with expression (3.9), expression (3.13) allows for the effect of the axial force. This is because the latter expression is derived from expression (3.7) which links the shear force (V) at location 2 with the compressive-force (F_c) developing at the same location due to flexure which may or may not combine with axial force.

(c) *Verification of methods of calculation*

It is interesting to note that the failure criteria proposed in the preceding section are dependent on a single material parameter, the tensile strength of concrete f_t, which must be divided by the code specified safety factor for concrete when the criteria are applied in practical design. The verification of their validity may be based on a comparison of the values of load-carrying capacity (expressed as shear force and bending moment) calculated from expressions (3.9) or (3.11) and (3.13) with published experimental values obtained from tests on beam-like elements without transverse reinforcement. Typical graphical representations of the calculated and experimental values expressing the variation of M_u/M_f with a_v/d for the case of $N = 0$ (extracted from Ref. [6]) are shown in Fig. 3.11 through to 3.13, with the figures also including predicted and experimental values for type III behaviour which forms the subject of Sect. 3.3.3.

Trends of behaviour. From the figures, it can be seen that the predicted variations of M_u/M_f with a_v/d provide a realistic representation of the experimentally established trends of beam behaviour, which, as indicated by the experimental results shown in Fig. 3.11 [8–10], appear to be affected by both ρ and f_c, with these effects being more clearly described in in Figs. 3.12 and 3.13, respectively.

Effect of ρ. Figure 3.12 shows the variation of M_u/M_f with a_v/d for the RC beams, with $f_c \approx 26$ MPa and ρ varying between 0.5 and 2.8 %, tested by Kani [8] under two-point loading. The figure indicates that the predictions provide a realistic description of the experimentally-established behaviour which is characterised by an increase in M_u/M_f with decreasing ρ. It should be noted that, in contrast with type III behaviour (i.e. for $1 \le a_v/d < 2.5$) for which M_u/M_f is always dependent on ρ (see Sect. 3.3.3), for type II behaviour (i.e. for $a_v/d \ge 2.5$) the dependence of ρ on M_u/M_f exists only when failure occurs due to the loss of bond between concrete and the flexural reinforcement, as indicated in expression (3.13), where $V_{II,2}$, as a function of F_c, is dependent on ρ.

Effect of f_c. Figure 3.13 shows typical variations of M_u/M_f with a_v/d for the RC beams, with $\rho = 1.88$ % and f_c varying between approximately 17 and 35 MPa tested by Kani under two-point loading [8]. The figure shows that the experimentally-established values of M_u/M_f increase with decreasing f_c and that, as for the

Fig. 3.11 Variation of $M_u/$ M_f with a_v/d for the RC beams, with f_c varying between approximately 26 and 39 MPa and ρ varying between 0.5 and 2.8 %, tested by various investigators, under two-point loading (suffixes C and E denote calculated and experimental values respectively)

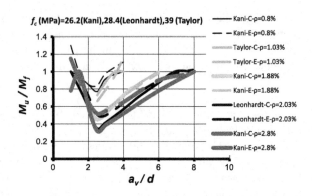

Fig. 3.12 Variation of $M_u/$ M_f with a_v/d for RC beams with $f_c \approx 26$ MPa and ρ varying between 0.5 and 2.8 %, tested by Kani [8] under two-point loading (prefixes C and E denote calculated and experimental values respectively)

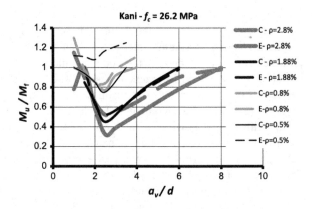

Fig. 3.13 Variation of $M_u/$ M_f with a_v/d for the RC beams, with $\rho = 1.88$ % and f_c varying between approximately 17 and 35 MPa, tested by Kani [8] under two-point loading (prefixes C and E calculated and experimental values respectively)

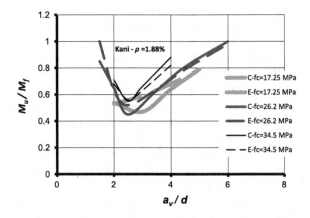

case of the effect of ρ, their predicted counterparts provide a realistic description of this trend of behaviour. The dependence of load-carrying capacity (expressed in the form of M_u/M_f) on f_c, is described by the failure criteria proposed as these link non-flexural types of failure with the development of transverse tensile stresses within the compressive zone of beam-like elements.

Typical results of a parametric study of the effect of axial compression on the failure load of beam-like elements exhibiting type II behaviour are presented in Table 3.2 extracted from Ref. [7]. The results were obtained from tests on knee frames, a schematic representation of which is provided in Fig. 3.14. The vertical load applied at the specimen ends combined with the 45° leg inclination resulted in the development of an axial compression (N) numerically equal to the shear force (V) within the frame legs.

The table includes experimentally established values of load-carrying capacity expressed in the form of "shear" capacity (V_{EXP}) and corresponding axial compression (N_{EXP}), together with the values predicted through the use of the proposed expressions (V_{PRE}) corresponding to N_{EXP}. The values (V_{ACI}, V_{EC2}) calculated through the use of the formulae incorporated in ACI310 [11] and EC2 [3] are also included in the tables for purposes of comparison. The table also contains the geometric characteristics and material properties required for assessing the specimens' load-carrying capacity, with full details of the specimens tested being provided in the relevant publication cited in the table caption.

From the table, it can be seen that the proposed expressions provide a close fit to the experimental values for all values of N. In fact, the fit provided by the proposed expressions—developed, as already discussed in the preceding section, from first principles without the need of calibration through the use of experimental data—is at least as close as that of the code formulae, which are empirical in nature and have been derived by regression analysis of experimental data, those of the table included.

3.3.3   Type III Behaviour

In contrast with type II behaviour, the brittle failure characterising type III behaviour (encompassing, approximately, the range of a_v/d between 1 and 2.5) is a flexural failure mode which, as discussed in Sect. 2.3, occurs due to failure of concrete in the compressive zone (the depth of which decreases considerably due to the deep penetration of the inclined crack that forms within the shear span) before yielding of the tension reinforcement. Since, as indicated in Fig. 2.9, the bending moment corresponding to load-carrying capacity, M_{III}, varies linearly with a_v/d for values of the latter between 1 and 2.5, the value M_{III} corresponding to any value of a_v/d between 1 and 2.5 may be easily accessed by linear interpolation between the values of M_{III} corresponding to $a_v/d = 1$ ($M_{III} = M_f$) and $a_v/d = 2.5$ ($M_{III} = M_{II}^{(2.5d)}$), i.e.

$$M_{III} = M_{II}^{(2.5d)} + \left(M_f - M_{II}^{(2.5d)} \right) (2.5d - a_v)/(1.5d) \qquad (3.14)$$

Where $M_{II}^{(2.5d)} = (2.5d)min(V_{II,1}, V_{II,2})$, with $V_{II,1}$ and $V_{II,2}$ resulting from expressions (3.9), or (3.11), and (3.13), respectively, is the value of M_{III} corresponding to $a_v/d = 2.5$, and M_f is the flexural capacity (for $a_v/d = 1$, $M_f = M_{III} = M_{IV}$).

Table 3.2 Calculated and experimental values of shear capacity of knee frames with type II behaviour ($2.5 < a_v/d$) [28]

Spec	$\rho\%$	f_y MPa	f_c MPa	N_{EXP} kN	V_{EXP} kN	V_{ACI} kN	V_{ACI}/V_{EXP}	V_{EC2} kN	V_{EC2}/V_{EXP}	V_{PRE} kN	V_{PRE}/V_{EXP}
Width (b) = 304.8 mm; height (h) = 406.4 mm; effective depth (d) = 368.3 mm; shear span (a_v) = 980.44 mm											
Non-flexural failure											
F38B2	1.91	374	12.4	113	113	70	0.62	116	1.03	80	0.71
F38E2	0.5	388	14.1	92	92	74	0.8	88	0.96	84	0.91
F38B4	1.84	386	31.4	173	173	115	0.66	162	0.93	149	0.86
F38D4	1.32	368	26.9	169	169	106	0.63	138	0.82	133	0.79
F38E4	0.91	368	32.1	148	148	115	0.78	128	0.87	151	1.02
F38A6	2.91	379	45.6	236	236	143	0.61	211	0.89	193	0.82
F38B6	1.81	379	41.6	198	198	134	0.68	175	0.88	191	0.96
shear span (a_v) = 1412.24 mm											
F55B2	1.88	375	11.9	94	94	68	0.72	112	1.19	77	0.82
F55E2	0.52	446	13.8	79	79	72	0.91	78	0.99	82	1.04
F55A4	1.98	406	26.4	151	151	104	0.69	152	1.01	128	0.85
F55B4	1.81	384	29.5	126	126	109	0.87	149	1.18	141	1.12
F55D4	1.45	432	25.6	126	126	101	0.8	134	1.07	126	1
F55E4	0.94	423	28.3	130	130	107	0.82	122	0.94	140	1.07
F55A6	3.32	379	42.1	189	189	134	0.71	208	1.1	177	0.94
F55B6	1.88	376	43.7	142	142	133	0.94	172	1.21	189	1.33
shear span (a_v) = 1778 mm											
F70B2	1.91	383	14.4	91	91	77	0.85	118	1.31	74	0.81
F70A4	2.16	376	29	142	142	111	0.78	159	1.12	134	0.95
F70A6	3.34	355	38.7	173	173	128	0.74	201	1.16	150	0.87
shear span (a_v) = 2133.6 mm											
F84B4	1.84	381	29.7	131	131	109	0.83	150	1.15	140	1.07
Mean values							**0.76**		**1.04**		**0.94**
Standard deviations							**0.1**		**0.14**		**0.15**
Flexural failure											
shear span (a_v) = 2870.2 mm											
F113B4	1.87	37.6	26	96	96					100	1.04

The use of expression (3.14) has been found to produce realistic predictions of structural behaviour which, as indicated in Figs. 3.11, 3.12 and 3.13, provide a close fit to the experimentally-established values of load-carrying capacity.

3.3.4 Type IV Behaviour

As described in Sect. 2.3, although the load-carrying capacity of beams with type IV behaviour corresponds to flexural capacity, loss of load-carrying capacity may be attributable to either failure of uncracked concrete in compression within the middle portion of the beam, i.e. failure of the horizontal element of the 'frame' of

Fig. 3.14 Schematic representation of knee specimens

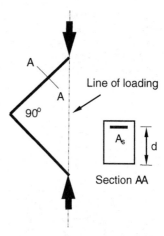

Fig. 3.15 Geometric characteristics and internal actions of 'frame' model of beam with type IV behaviour

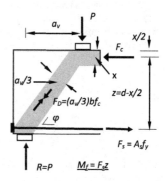

the proposed model, or failure in compression of the uncracked end portion of the beam, i.e. failure of the inclined leg of the 'frame' of the proposed model. In the former case, failure may be said to be ductile (resembling flexural failure), while in the latter case, it is brittle (resembling uniaxial compression failure). As brittle failure is undesirable, the prediction of only the load-carrying capacity is not adequate for design purposes; the prediction must also include the mode of failure. A simple method that can be used for predicting both load-carrying capacity and mode of failure for the case of two-point loading may be as follows (see also Fig. 3.15) [12].

Assuming the beam geometric characteristics and longitudinal reinforcement are given, the flexural capacity (M_f) is calculated as described in Fig. 3.3 with the load-carrying (P_f) capacity corresponding to it being easily obtained from the moment equilibrium of the free body in Fig. 3.15, i.e.

$$P_f = M_f/a_v \tag{3.15}$$

On the other hand, the load-carrying capacity (P_D) corresponding to the strength of the inclined leg of the 'frame' will be equal to the vertical component of the load (F_D) that can be carried by this leg. As indicated in Fig. 3.15, F_D is easily calculated by taking the depth of the leg equal to $a_v/3$ as recommended in Ref. [13], i.e.

Fig. 3.16 Correlation between predictions and experimental values reported in [18] and [19] for the load-carrying capacity of beams with type IV behaviour

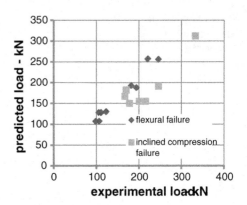

Fig. 3.17 Correlation between predictions and experimental values reported in [20] for the load-carrying capacity of beams with type IV behaviour

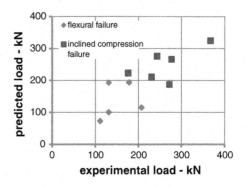

Fig. 3.18 Correlation between predictions and experimental values reported in [21] for the load-carrying capacity of beams with type IV behaviour

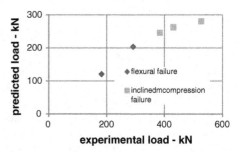

$$P_D = F_D z / \left(z^2 + a_v^2 \right)^{1/2} \tag{3.16}$$

where $F_D = (a_v/3)\, b f_c$

Therefore, the load-carrying capacity of a beam of type IV behaviour will be

$$P_u = \min \left(P_f, P_D \right) \tag{3.17}$$

An indication of the validity of the proposed method is provided in Figs. 3.16, 3.17, 3.18, 3.19 and 3.20 which show the relationship between predicted and experimental values of the load-carrying capacity, together with predictions of the mode

Fig. 3.19 Correlation
between predictions and
experimental values reported
in [20] for the load-carrying
capacity of beams with type
IV behaviour

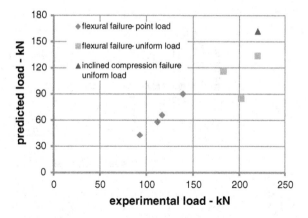

Fig. 3.20 Correlation
between predictions and
experimental values reported
in [22] for the load-carrying
capacity of beams with type
IV behaviour

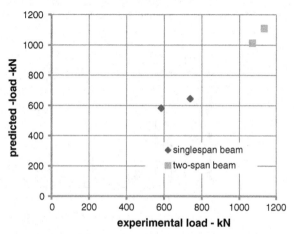

of failure. From the figures, it becomes apparent that the predictions of the proposed method are satisfactory for all cases investigated, in spite of the simplifications incorporated in the proposed criterion expressed by expressions (3.15) to (3.17).

3.4 Comparison of Predictions of Proposed and Code Adopted Criteria

All predicted values are expressed in a normalised form as the ratio of the experimental (V_E) to the calculated values of shear force at failure and presented in a bar chart form in Figs. 3.21, 3.22, 3.23, 3.24, 3.25, 3.27, 3.28, 3.29, 3.30, and 3.31. The calculated values include those predicted by the proposed method (V_C), ACI (Eq. (11.5) in the code [11]) (V_{ACI}) and EC2 (Eq. 6.2 in the code [3]) (V_{EC2}). The figures also provide the designation of the specimens through which their design details may be obtained from the references cited in the figure captions.

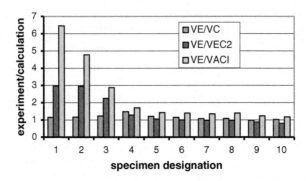

Fig. 3.21 Ratios of experimental (V_E) to calculated (through the proposed (V_C), EC2 (V_{EC2}), and ACI (V_{ACI}) methods) values of load-carrying capacity of RC beams tested under two-point loading by Leonhardt and Walther [9]. (Average and standard deviation values of ratios for proposed, EC2 and ACI methods: 1.15 and 0.14, 1.5 and 0.74, 1.92 and 1.19, respectively)

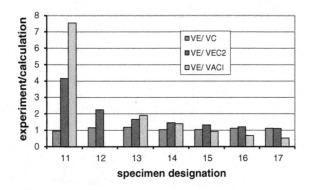

Fig. 3.22 Ratios of experimental (V_E) to calculated (through the proposed (V_C), EC2 (V_{EC2}), and ACI (V_{ACI}) methods) values of load-carrying capacity of RC beams tested under uniform loading by Leonhardt and Walther [9] (Average and standard deviation values of ratios for proposed, EC2 and ACI methods: 1.09 and 0.08, 1.88 and 1.07, 2.32 and 2.48, respectively)

Figures 3.21, 3.22, 3.23, 3.24, 3.25, 3.27, 3.28, 3.29, 3.30, and 3.31 indicate that the proposed failure criteria provide a fit to the experimental values of load-carrying capacity significantly closer than that of the code formulae. The information provided in Figs. 3.21, 3.22, 3.23, 3.24 and 3.25 refers to structural elements made of normal-strength concrete, whereas similar information for structural elements made of high-strength concrete is included in Figs. 3.27, 3.28, 3.29 and 3.30; finally, Fig. 3.31 provides an indication of the effect on load-carrying capacity of the size of the element and the provision of nominal reinforcement.

For the case of normal-strength concrete (see Figs. 3.21, 3.22, 3.23 and 3.24), it may be noted that the values of load-carrying capacity (V_C) calculated through the proposed method correlate closely with their experimental counterparts (V_E), with the mean values of the ratio V_E/V_C varying between 0.98 and 1.15, and exhibiting

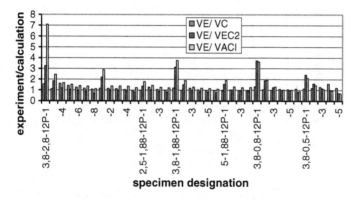

Fig. 3.23 Ratios of experimental (V_E) to calculated (through the proposed (V_C), EC2 (V_{EC2}), and ACI (V_{ACI}) methods) values of load-carrying capacity of RC beams tested under two-point loading by Kani [8] (Average and standard deviation values of ratios for proposed, EC2 and ACI methods: 1.12 and 0.16, 1.15 and 0.5, 1.46 and 0.59, respectively)

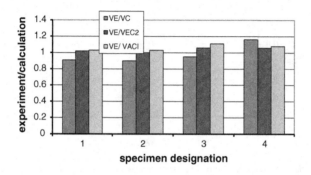

Fig. 3.24 Ratios of experimental (V_E) to calculated (through the proposed (V_C), EC2 (V_{EC2}), and ACI (V_{ACI}) methods) values of load-carrying capacity of RC beams tested under two-point loading by Taylor [10] (Average and standard deviation values of ratios for proposed, EC2 and ACI methods: 0.98 and 0.12, 1.04 and 0.03, 1.06 and 0.04, respectively)

a standard deviation ranging between 0.08 and 0.16. This range of values of the standard deviation lies well within the scatter of the experimentally established values of the direct tensile strength (f_t) of concrete. In fact, from Eq. (3.2a) it appears that the values of f_t exhibit a standard deviation of around 0.32 MPa, which is twice as large as that of the V_E/V_C ratio. On the other hand, the code predictions (V_{ACI}: ACI predicted values; V_{EC2}: EC2 predicted values) exhibit a considerable deviation from the experimental values which is reflected in both the mean value and the standard deviation of the V_E/V_{EC2} and V_E/V_{ACI} ratios.

In contrast with Figs. 3.21, 3.22, 3.23, 3.24, and 3.25 indicates that the proposed failure criteria overestimate the load-carrying capacity of the beams with $a_v/d = 1.5$ (beams 2, 4, and U) tested by Brown et al. [14] by a margin of just over 30 %. However, this deviation is compatible with the scatter exhibited by the

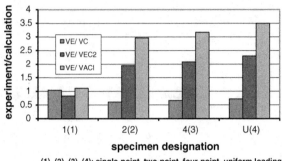

Fig. 3.25 Ratios of experimental (V_E) to calculated (through the proposed (V_C), EC2 (V_{EC2}), and ACI (V_{ACI}) methods) values of load-carrying capacity of RC beams tested under various loading regimes by Brown et al. [14] (Average and standard deviation values of ratios for proposed, EC2 and ACI methods: 0.76 and 0.19, 1.79 and 0.66, 2.69 and 1.07, respectively)

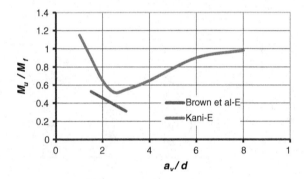

Fig. 3.26 Comparison of variations of M_u/M_f with a_v/d established for the RC beams, by Brown et al. [14] ($\rho = 3.07$ % and $f_c = 26.8$ MPa) and Kani [6] ($\rho = 2.88$ % and $f_c = 26.2$ MPa) (suffix E denotes experimental values) (see color figure online)

experimentally established values of f_t which may also be the cause underlying the deviation of the $M_u/M_f - a_v/d$ curve constructed from the experimental results obtained by Brown et al. [14] from that constructed from the experimental results obtained by Kani [8] for RC beams with similar f_c and ρ (see Fig. 3.26). In contrast with the proposed criteria, the code predictions underestimate load-carrying capacity by an amount ranging from about 80 to 170 %.

For the case of high-strength concrete, Figs. 3.27, 3.28, 3.29 and 3.30 indicate that the mean values of the V_E/V_C ratio are similar to those for normal-strength concrete, ranging between approximately 1.0 and 1.13. However, these values are characterised by a significantly larger standard deviation ranging between approximately 0.19 and 0.35, reflecting the large scatter of the experimental results, particularly for $a_v/d < 3$. This large scatter cannot be entirely attributed to the nature of high-strength concrete, since in programmes where similar tests were carried out on specimens made of normal-strength concrete, the results obtained appear

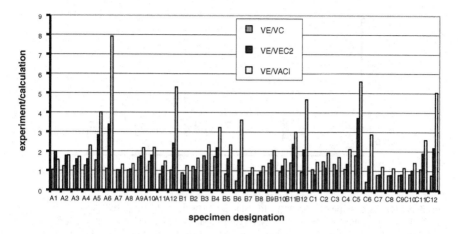

Fig. 3.27 Ratios of experimental (V_E) to calculated (through the proposed (V_C), EC2 (V_{EC2}), and ACI (V_{ACI}) methods) values of load-carrying capacity of RC beams tested under two-point loading by Ahmad and Lu [23] (Average and standard deviation values of ratios for proposed, EC2 and ACI methods: 1.1 and 0.5, 1.46 and 0.67, 1.97 and 0.97, respectively)

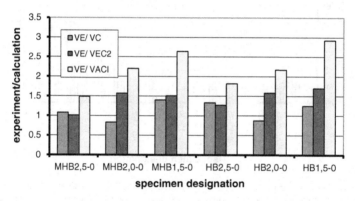

Fig. 3.28 Ratios of experimental (V_E) to calculated (through the proposed (V_C), EC2 (V_{EC2}), and ACI (V_{ACI}) methods) values of load-carrying capacity of RC beams tested under a single-point loading by Shin et al. [24] (Average and standard deviation values of ratios for proposed, EC2 and ACI methods: 1.13 and 0.24, 1.44 and 0.25, 2.21 and 0.52, respectively)

to exhibit a similar scatter (see, for example, Fig. 3.30). As for the case of normal-strength concrete, the code predictions are found to considerably underestimate load-carrying capacity.

Size effects. Figure 3.31 provides the ratios of the experimental to the calculated values of load-carrying capacity of specimens of various sizes, tested in an attempt to establish the effect of the element size on the calculated value of load-carrying capacity. The proposed failure criteria have been found to provide realistic predictions of load-carrying capacity for beams with a cross-section depth d up to about 750 mm; for larger values of d, the deviation of the calculated value

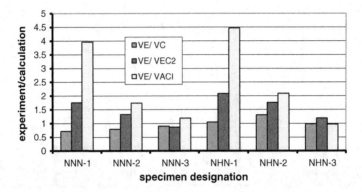

Fig. 3.29 Ratios of experimental (V_E) to calculated (through the proposed (V_C), EC2 (V_{EC2}), and ACI (V_{ACI}) methods) values of load-carrying capacity of RC beams tested under a single-point loading by Xie et al. [25] (Average and standard deviation values of ratios for proposed, EC2 and ACI methods: 1.0 and 0.19, 1.29 and 0.37, 1.5 and 0.51, respectively)

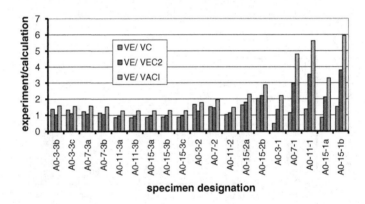

Fig. 3.30 Ratios of experimental (V_E) to calculated (through the proposed (V_C), EC2 (V_{EC2}), and ACI (V_{ACI}) methods) values of load-carrying capacity of RC beams tested under a single-point loading by Mphonde and Frantz [26] (Average and standard deviation values of ratios for proposed, EC2 and ACI methods: 1.19 and 0.38, 1.61 and 0.91, 2.6 and 1.5, respectively)

from its experimental counterpart appears to increase sharply (see calculated values for beams tested by Angelakos et al. [15] and beam BN100 tested by Collins and Kuchma [16]). However, such deviations disappear in the presence of nominal transverse reinforcement sufficient to sustain a tensile stress of the order of 0.5 MPa (see also calculated values for beams with nominal reinforcement tested by Angelakos et al. [15] (beams with suffix M)) as suggested elsewhere [17].

As regards the code formulae, one would expect them to yield closer predictions of load-carrying capacity as these, in contrast with the proposed criteria, allow for size effects. Nevertheless, the predicted values are not as close to their experimental counterparts as those obtained from the proposed criteria.

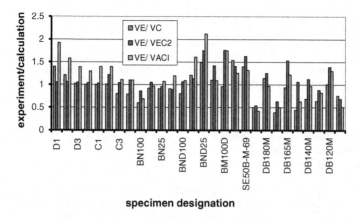

specimen designation

Fig. 3.31 Ratios of experimental (V_E) to calculated (through the proposed (V_C), EC2 (V_{EC2}), and ACI (V_{ACI}) methods) values of load-carrying capacity of RC beams of various sizes tested by Leonhardt and Walther [9] (beams with prefixes D and C), Collins and Kuchma [16] [beams with prefixes BN, BND, BM (with nominal stirrups), and SE (with nominal stirrups)], Angelakos et al. [15] (beams with a prefix DB, with those with a suffix M with nominal stirrups), and Lubell et al. [27] (beam AT1) (Average and standard deviation values of ratios for proposed, EC2 and ACI methods: 1.03 and 0.25, 1.13 and 0.31, 1.16 and 0.41, respectively)

3.5 Concluding Remarks

The CFP concept leads to the development of simple criteria for the non-flexural types of failure of RC beam-like elements without transverse reinforcement which were derived from first principles without the need of calibration through the use of experimental data on structural concrete behaviour. These criteria have been found not only to provide a realistic description of the causes of failure, but also to fit a wide range of published experimental data considerably more closely than the formulae adopted by current codes for assessing shear capacity.

References

1. Kotsovos GM (2011) Assessment of the flexural capacity of RC beam/column elements allowing for 3D effects. Eng Struct 33(10):2772–2780
2. EN 1998-1 (2004) Eurocode 8: design of structures for earthquake resistance—part 1: general rules, seismic actions and rules for buildings
3. EN 1992-1 (2004) Eurocode 2: design of concrete structures—part 1-1: general rules and rules for buildings
4. Kotsovos MD (1982) A fundamental explanation of the behaviour of reinforced concrete beams in flexure based on the properties of concrete under multiaxial stress. Mater Struct RILEM 15(90):529–537
5. Eibl J (ed) (1994/96) Concrete structures Euro-design handbook. Ernst and Sohn, Berlin, pp 764
6. Kotsovos GM, Kotsovos MD (2008) Criteria for structural failure of RC beams without transverse reinforcement. Struct Eng 86(23/24):55–61

7. Kotsovos GM, Kotsovos MD (forthcoming) Effect of axial compression on shear capacity of linear RC members. Mag Concr Res
8. Kani GNJ (1966) Basic facts concerning shear failure. J ACI 63(6):675–691
9. Leonhardt F, Wather R (1961) The Stuttgart shear tests, 1961: contributions to the treatment of the problems of shear in reinforced concrete construction. Translation No. 111 (A translation of the articles that appeared in beton-und Stahlbetonbau 56(12); Cement Concr Assoc 57(2, 3, 6, 7, 8), (1962)
10. Taylor R (1960) Some shear tests on reinforced concrete beams without shear reinforcement. Mag Concr Res 12(35):145–154
11. American Concrete Institute (2002) Building code requirements for structural concrete (ACI 318-02) and Commentary (ACI 318R-02)
12. Kong FK (ed) (1990) Reinforced concrete deep beams. Blackie and Van Nostrand Reinhold, New York, pp 288
13. The Institution of Structural Engineers (1978) Design and detailing of concrete structures for fire resistance. Interim guidance by a joint committee of the Institution of Structural Engineers and the Concrete Society. The Institution of Structural Engineers, pp 59
14. Brown MD, Bayrak O, Jirsa JO (2006) Design of shear based on loading conditions. ACI Struct J 103(4):541–550
15. Angelakos D, Bentz EC, Collins MP (2001) Effect of concrete strength and minimum stirrups on shear strength of large members. ACI Struct J 98(3):290–300
16. Collins MP, Kuchma D (1999) How safe are our large, lightly reinforced concrete beams, slabs, and footings. ACI Struct J 96(4):482–490
17. Kotsovos MD, Pavlovic MN (1997) Size effects in structural concrete: a numerical experiment. Comput Struct 64(1–4):285–295
18. Rawdon de Paive HA, Siess CP (1965) Strength and behaviour of deep beams in hear. J Struct Div Proc ASCE 91(ST5):19–41
19. Smith KN, Vantsiotis AS (1982) Shear strength of deep beams. ACI J 79(3):201–213
20. Ramakrishna V, Ananthanarayana Y (1968) Ultimate strength of deep beams in shear. ACI J 65(2):87–98
21. Kong FK, Robins PJ, Cole DF (1970) Web reinforcement effects in deep beams. ACI J 67(12):1010–1017
22. Rogowski DM, MacGregor JG, Ong SY (1986) Tests of reinforced concrete deep beams. ACI J 83(4):614–623
23. Ahmad SH, Lue DM (1987) Flexure-shear interaction of reinforced high-strength concrete beams. ACI Struct J 84(4):330–341
24. Shin S-W, Lee K-S, Moon J-I, Ghosh SK (1999) Shear strength of reinforced high-strength concrete beams with shear span-to-depth ratios between 1.5 and 2.5. ACI Struct J 96(4):549–556
25. Xie Y, Ahmad SH, Yu T, Hino S, Chung W (1994) Shear ductility of reinforced beams with normal and high-strength concrete. ACI Struct J 91(2):140–9
26. Mphonde G, Frantz GC (1984) Shear tests of high- and low-strength concrete beams without stirrups. ACI Struct J 81(4):350–357
27. Lubell A, Sherwood T, Bentz E, Collins MP (2004) Safe shear design of large, wide beams: adding shear reinforcement is recommended. Concr Int 26(1):67–78
28. Morrow J, Viest IM (1957) Shear strength of reinforced concrete frame members without web reinforcement. ACI J Proc 53:833–886
29. Kotsovos GM, Kotsovos DM, Kotsovos MD, Kounadis A (2011) Seismic design of structural concrete walls: an attempt to reduce reinforcement congestion. Mag Concr Res 63(4):235–245
30. Kotsovos GM (2011) Seismic design of RC beam-column structural elements. Mag Concr Res 63(7):527–537

Chapter 4
Design of Simply Supported Beams

4.1 Introduction

In this chapter, the physical model and failure criteria presented in Chap. 3 are used within the framework of current design philosophy for assessing the longitudinal and transverse reinforcement required for an RC beam to satisfy the requirements of current codes for load-carrying capacity and ductility.

4.2 Assessment of Longitudinal Reinforcement

For a cross section with given geometry (i.e. shape and dimensions) and material characteristics, the longitudinal reinforcement (both reinforcement in tension A_s and reinforcement in compression A'_s) required for the cross section to be capable of sustaining the action of a specified design bending moment M_d can be assessed through the following step by step procedure which is based on the method of calculation of the cross section's flexural capacity M_f described in item (b) of Sect. 3.3.1:

(1) Set $A'_s = 0$ and $A_s = A_{s(o)} = M_d/(zf_y)$, where f_y is the yield stress of the steel and $z = 0.9d$.

(2) Calculate x from equation (a) in Fig. 4.1 (top), where $F_s = A_s f_y$ and $F_c = \sigma_a bx$ in which σ_a results from expression 3.3 by replacing f_t with its value obtained from one of expressions 3.2 or 3.2b depending on the value of f_c.

(3) Calculate M_f from equation (b) in Fig. 4.1 (top).

(4) Compare M_f with M_d:

  (4.1) If $0 \leq M_f - M_d \leq a$, where a is an acceptable small (positive) value, the values of A_s and A'_s adopted for assessing M_f are the required values. If $M_f - M_d > a$, return to step (1), adjust the value of A_s and repeat process.

M. D. Kotsovos, *Compressive Force-Path Method*, Engineering Materials,
DOI: 10.1007/978-3-319-00488-4_4, © Springer International Publishing Switzerland 2014

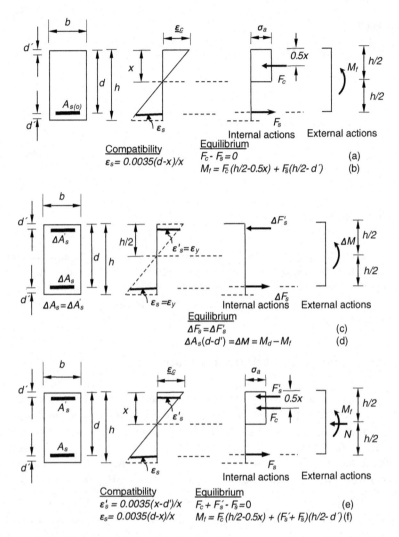

Fig. 4.1 Calculation of flexural capacity for $A'_s = 0$ and $A_s = A_{s(o)}$ (*top*), for $A'_s = A_s = \Delta A_s$ (*middle*) and for $A'_s = \Delta A_s$ and $A_s = A_{s(o)} + \Delta A_s$ (*bottom*)

(4.2) If $M_f - M_d \leq 0$, place additional amounts of reinforcement in tension ΔA_s and compression $\Delta A'_s = \Delta A_s$ with their geometric centres lying at distances d and d' from the extreme compressive fibre of the cross section, such that $\Delta A_s (d-d') = \Delta M = M_d - M_f$ [see Fig. 4.1 (middle)]; then, return to (2) and repeat the process by using $A'_s = \Delta A_s$ and $A_s = A_{s(o)} + \Delta A_s$ and taking into account the compatibility equations (see Fig. 4.1 (bottom)) until $0 \leq M_f - M_d \leq a$. (The latter condition is usually fulfilled with the first attempt).

4.3 Assessment of Transverse Reinforcement

For behaviour of types II and III, transverse reinforcement may be required in order to prevent failure from occurring before flexural capacity is exhausted. In all other cases, a nominal amount of reinforcement in the form of stirrups with a spacing not larger than $d/2$ is deemed sufficient for sustaining tensile stresses of the order of 0.5 MPa. (In most cases, such reinforcement is somewhat more conservative than the value specified by current codes).

4.3.1 Type II Behaviour

As discussed in Sect. 2.3, one of the modes of failure characterising beams of type II behaviour takes the form of near-horizontal splitting of the compressive zone of the beam in the region (marked with "1" in Fig. 3.7 of the preceding chapter) of the joint of the horizontal and inclined elements of the 'frame' model of the beam (i.e. the region where the path of the compressive stress resultant developing on account of bending changes direction). Such splitting is caused by the development of a tensile force which is essentially equivalent to the shear force acting in this region. (As explained in Sect. 2.2.5, this can also be understood by viewing the 'kink' in the stress path as giving rise to an orthogonal force bisecting the angle between the horizontal portion and the inclined leg of the frame: that such a force is tensile is evident from the fact that it tends to separate the compressive zone of the beam from the cracked region below it).

The maximum value of the shear force that can be sustained by concrete in the above region easily results from expression 3.9. When the value of the acting shear force becomes larger than the value resulting from expression 3.9, the mode of failure described above can be prevented through the use of transverse reinforcement in the form of stirrups. Following the current code reasoning, such reinforcement is placed in an amount sufficient to sustain the whole shear force corresponding to flexural capacity. Moreover, since as discussed in item (a) of Sect. 3.3.2, the tensile stresses developing in the region of the change in the force-path direction spread over a length equal to d on the either side of the cross section at a distance equal to $2.5d$ from the support, the reinforcement is uniformly distributed over the same length.

Therefore, in order to safeguard against failure due to horizontal splitting of the compressive zone occurring in location 1 (see Fig. 3.7) before flexural capacity is exhausted, the acting shear force at this location is taken equal to V_f, then the total amount of transverse reinforcement required to sustain V_f is given by

$$A_{sv,II1} = V_f/f_{yv} \qquad (4.1)$$

where f_{yv} is the yield stress of the reinforcement.

Such reinforcement (placed within a length equal to $2d$ spreading symmetrically about the cross section at a distance of $2.5d$ from the support) is considered to be fully effective when the stirrup spacing is not larger that $0.5d$.

As discussed in Sect. 2.2.5, for the case of a beam subjected to two point loading, the causes of failure characterising type II behaviour may also be associated with the loss of bond between concrete and the longitudinal reinforcement in tension in the region (marked with "2" in Fig. 3.7 of the preceding chapter) of the shear span adjacent to a point load (see item (d) in Sect. 2.2.5). (Such bond failure is only associated with type II behaviour and, say, two-point loading, since uniformly-distributed loading results in small shear forces throughout much of the central span: for more than two point loads, an equivalent uniformly-distributed loading can be assumed but, if in doubt, a check for shear can always be made for a (small) finite number of point loads). Figure 3.8 illustrates a portion of the beam between consecutive cracks, together with the internal actions developing on these sections before and after loss of bond.

The manner in which the loss of bond causes the change in the internal actions and leads to the development of transverse actions within the compressive zone of the element, as indicated by σ_t in Fig. 3.8 and discussed in Sect. 3.3.2, is fully described in Sect. 2.2.5 by reference to Fig. 2.8. An assessment of the value of transverse tensile stresses σ_t developing in the compressive zone, when the beam's flexural capacity is attained, can be made through the use of expression 3.13 by replacing $|f_t|$ with $|\sigma_t|$ and $V_{II,2}$ with V_f (the latter being the shear force at the location (marked with "2" in Fig. 3.7) corresponding to flexural capacity) and solving for $|\sigma_t|$; then

$$|\sigma_t| = f_c / \left[5\left(F_c/V_f - 1\right)\right] \tag{4.2}$$

The value of σ_t obtained from expression (4.2) is used to obtain the vertical and horizontal stress resultants per unit length of the beam within the region marked "2" in Fig. 3.7,

$$T_{II,2v} = \sigma_t b/2 \tag{4.3a}$$

$$T_{II,2h} = \sigma_t x/2 \tag{4.3b}$$

(the second term is divided by 2 since, as indicated in Fig. 3.8, the transverse tensile stresses due to the loss of bond develop within half the length of the element considered).

Therefore, the amount of transverse reinforcement required per unit length to sustain $T_{II,2v}$ and $T_{II,2h}$ will be equal to

$$A_{sv,II2v} = T_{II,2v}/f_{yv}; \quad A_{sv,II2h} = T_{II,2h}/f_{yv} \tag{4.4}$$

Such reinforcement is placed per unit length of the beam in the region marked "2" in Fig. 3.7; this region extends between the cross section of the maximum acting bending moment and the cross section at a distance of $2.5d$ from the nearest support.

Fig. 4.2 Internal actions on
the portion of an RC beam
extending to the cross section
through the tip of an inclined
crack for type III behaviour

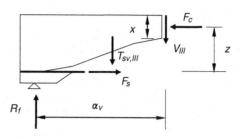

4.3.2   Type III Behaviour

As discussed in Sect. 2.3, failure characterising type III behaviour is attributable to
the reduced strength of the compressive zone of the uncracked portion of the beam
(i.e. the horizontal element of the 'frame' of the proposed model) adjacent to the
region of the change in the path of the compressive stress resultant (i.e. adjacent to
the region of the joint of the horizontal and inclined members of the 'frame'). This
strength reduction, which is due to the deep penetration of the inclined crack closest
to the support into the compressive zone (see Fig. 2.1), results in a reduction of the
beam's flexural capacity.

The beam's load-carrying capacity may be increased to the value corresponding
to the flexural capacity by uniformly distributing transverse reinforcement within
the whole length of the horizontal projection of the inclined leg of the frame (i.e.
of the horizontal projection of the inclined crack closest to the support [see Fig.
2.1)]. Figure 4.2 depicts the portion of the beam in Fig. 2.1 enclosed by its left-
hand end face, the deepest inclined crack closest to the support, and the cross-section of the horizontal element with the reduced strength. If it is assumed that the
transverse reinforcement is at yield, then the total force that can be sustained by
such reinforcement is $T_{sv,III} = A_{sv,III}f_{yv}$ which acts in the middle of the shear span
a_v of the portion considered. For the equilibrium of this portion,

$$R_f a_v - T_{sv,III}\,(a_v/2) - F_s z = 0 \qquad (4.5)$$

where $F_s\,z = V_{III}\,\alpha_v = M_{III}$ [i.e. the bending moment that can be sustained by the
beam in the absence of any transverse reinforcement; the value of M_{III} is obtained
from expression (3.14)], and $R_f\,a_v = M_f$ (i.e. the flexural capacity).

Replacing in expression (4.5) $R_f a_v$ with M_f, $F_s z$ with M_{III}, $T_{sv,III}$ with $A_{sv,III}f_{yv}$
and solving for A_{sv} yields

$$A_{sv,III} = 2(M_f - M_{III})/\,(a_v f_{yv}) \qquad (4.6)$$

4.4   Design Procedure

The design of a reinforced concrete beam involves, on the one hand, the selection of materials (concrete and steel) of a suitable quality, and, on the other hand,
the determination of the geometric characteristics (i.e. shape and dimensions) of

the member, inclusive the amount and location of reinforcement, required for the beam to have a given load-carrying capacity and ductility. However, the selection of the quality of the materials is independent of the proposed methodology, since the underlying theory is valid for the whole range of material qualities available to date for practical applications. Also, out of the geometric characteristics, the span of the beam may be considered as known, since it results directly from the overall structural configuration adopted. Hence, the design of a simply-supported reinforced concrete beam involves essentially the determination of the cross-sectional characteristics (i.e. shape and dimensions) of the member together with the amount and location of the reinforcement.

Bearing in mind the above, the information presented in the preceding chapter and Sects. 4.2 and 4.3 of the present chapter may be incorporated into the following design procedure which comprises six steps:

(a) *Preliminary assessment of geometric characteristics.* This may be carried out by following current design practice as described in Ref. 1.1. For example, with the exemption of deep beams (i.e. beams characterised by IV behaviour), the cross-sectional depth (d) is taken approximately equal to $L/12$, where L is the beam span, while the web width (b_w) of the cross-section is given a value between $d/3$ and $2d/3$. (For a rectangular cross-section, $b = b_w$). For the case of deep beams, the beam depth is such as to satisfy the condition $L/2d < 1$, which defines deep-beam behaviour, while the width may be taken initially to be equal to $L/24$.

(b) *Calculation of design bending moment and shear force.* With the applied load and the beam span known, the bending moment and shear-force diagrams may be easily constructed. The design bending moment and shear force at 'critical' cross-sections are obtained from these diagrams.

(c) *Assessment of longitudinal (flexural) reinforcement.* If it is assumed that the flexural capacity (M_f) is at least equal to the design bending moment (M_d), the amount (A_s) of the longitudinal reinforcement required is calculated as described in Sect. 4.2. (It should be noted from Fig. 3.3 that the calculation of A_s is preceded by the calculation of the depth x of the beam's compressive zone (i.e. the depth of the horizontal element of the model 'frame').

(d) *Construction of physical model.* As indicated in Fig. 3.1, the shape of the model is essentially given; only the position of the joint of the horizontal and inclined members of the 'frame' must be determined. Figure 3.1 defines the above position for the cases of the two-point loading (which also describes the case of single-point loading if the two point loads have a common point of application, in which instance the 'horizontal' member shrinks to the joint of the two inclined legs) and uniformly-distributed loading. For a number of point loads larger than two, point-loading may be considered equivalent to uniformly distributed loading with the same total load.

(e) *Determination of type of beam behaviour.* From Figs. 2.9 and 3.1, the type of behaviour may be determined by the value of a_v/d or L/d depending on the type of the applied load.

(f) *Calculation of transverse reinforcement.* Transverse reinforcement is required only for behaviour of types II and III. For all other cases, it is sufficient to provide nominal reinforcement in the form of stirrups with a spacing not larger than $d/2$ and capable of sustaining tensile stresses of the order of 0.5 MPa. For the case of type II behaviour, the method of calculation of transverse reinforcement (required for all types of loading at the location of the joint of the horizontal and inclined elements of the model's frame) is described in Sect. 4.3.1. This section also describes the method of calculation of the additional reinforcement that may be required to prevent horizontal splitting of the compressive zone for the case of point-loading. Throughout the remainder of the beam, nominal reinforcement is provided as for the cases I and IV. For type III behaviour, the calculation of transverse reinforcement is carried out as described in Sect. 4.3.2.

4.5 Design Examples

In what follows, the examples show instances of design of simply-supported beams exhibiting types of behaviour II, III and IV, with the design of such beams implicitly including the case of type I behaviour, since its aim is to safeguard the latter type of behaviour. The geometric characteristics of the beams' cross section are assessed as discussed in the preceding section; for the cases of type II and III behaviour, attention is primarily focused on the assessment of the transverse reinforcement required to safeguard against brittle failure.

4.5.1 Beam Under Uniformly-Distributed Loading

The beam illustrated in Fig. 4.3 has a rectangular cross-section and a span $L = 6,000$ mm. It is constructed with concrete with a uniaxial cylinder compressive strength $f_c = 30$ MPa and steel bars with a yield stress $f_y = 500$ MPa, for both longitudinal and transverse reinforcement.

Fig. 4.3 Geometric characteristics (*top*) and physical model (*bottom*) of simply-supported beam exhibiting type I behaviour under uniformly distributed load

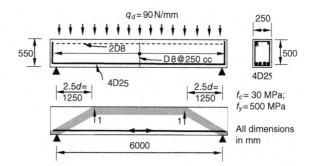

(a) *Assessment of cross-sectional geometric characteristics*

As discussed in item (a) of the preceding section, d is taken equal to approximately $L/12$, i.e. $d \approx L/12 = 6{,}000/12 = 500$ mm, and b is given a value between $d/2$ and $2d/3$, say $b = d/2 = 250$ mm. For $q_d = 90$ N/mm, the design bending moment $M_d = q_d L^2/8 = 90 \times 6{,}000^2/8 = 405 \times 10^6$ Nmm; assuming $z = 0.9d = 450$ mm, the amount of longitudinal reinforcement required for the beam to be capable of sustaining the design bending moment is assessed as discussed in item (1) of Sect. 4.2, i.e. $A_s = M_d/(zf_y) = 405 \times 10^6/(450 \times 500) = 1{,}800$ mm^2. Such a value of A_s is equivalent to four 25 mm diameter bars (4D25 $\approx$ 1,964 mm^2).

(b) *Verification of adequacy of A_s*

For 4D25, the tensile force sustained by the steel bars is $F_s = A_s f_y = 1{,}964 \times 500 = 982000$ N. The equilibrium condition $F_c = F_s$ [see Fig. 4.1 (top)] yields $x = F_s/(\sigma_a b)$. From expression 3.3, in which $f_t = 2.37$ MPa is obtained from expression 3.2a, $\sigma_a = 41.84$ MPa and thus $x = 982{,}000/(41.84 \times 250) \approx 94$ mm. Hence, the lever arm of the couple of the longitudinal internal forces is $z = d - x/2 = 500 - 94/2 = 453$ mm and the moment of the couple yields the beam's flexural capacity $M_f = F_s z = 982{,}000 \times 453 \approx 445 \times 10^6$ N mm $> M_d = 405 \times 10^6$ Nmm. Therefore, the reinforcement provided is sufficient for the beam to sustain the design bending moment.

(c) *Physical model*

For the case of a simply-supported beam under uniformly distributed loading, the position of the joint of the horizontal and inclined elements of the model 'frame' depends on L/d (see Fig. 3.1). Since $L/d = 6{,}000/500 = 12 > 8$, the beam is characterised by type of behaviour I or II and, hence, the distance of the joint from the support is equal to $2.5d = 2.5 \times 500 = 1{,}250$ mm [see Fig. 4.3 (bottom)].

(d) *Checking for non-flexural failure*

For the case of beams exhibiting type II behaviour under uniformly-distributed loading, a non-flexural type of failure may only occur due to the development of transverse tension in the region of the joint of the horizontal and inclined elements of the 'frame' indicated as location 1 in Fig. 4.3 (bottom). This tensile force, which is numerically equal to the shear force ($V_{II,1}$) developing at this location, cannot exceed a value $T_{II,1} = 0.5 \times 2.37 \times 250 \times 500 = 148125$ N obtained from expression (3.8).

The load-carrying capacity of the beam corresponding to M_f is $q_f = 8\,M_f L^2 = 8 \times 445 \times 10^6/6{,}000^2 \approx 99$ N/mm and the shear force developing at location 1 is $V_f = q_f L/2 - 2.5\,d\,q_f = 99 \times 6{,}000/2 - 2.5 \times 500 \times 99 = 173250$ N $> V_{II,1} = T_{II,1} = 148125$ N. Hence, the beam is not capable of sustaining the tensile force developing at location 1 when the beam's flexural capacity is attained.

(e) *Transverse reinforcement*

In view of the above, there is a need for transverse reinforcement which, as discussed in Sect. 4.3.1, should be capable of sustaining the total transverse tensile force of 173250 N developing within the region of location 1. The total amount of such reinforcement is $A_{sv,III} = 173,250/500 = 346.5$ mm^2 which is placed within a length $2d = 2 \times 500 = 1,000$ mm, symmetrically with respect to location 1, the latter being at a distance equal to 2.5 $d = 1,250$ mm from the support closest to it. This reinforcement is equivalent to four two-legged 8 mm diameter stirrups at a spacing of 290 mm ($=402.12$ mm$^2 > 346.5$ mm$^2) > d/2 = 250$ mm. Since, the calculated spacing is larger than the maximum allowed value of $d/2 = 250$ mm, the specified stirrups are placed at a spacing of 250 mm throughout the beam span in order to satisfy the requirement of nominal reinforcement. (It should be noted that the specified stirrups are capable of sustaining transverse tensile stresses of 0.8 MPa which is larger than the nominal value of 0.5 MPa).

4.5.2 Beam Under Two-Point Loading with $a_v/d = 4$

In what follows, the beam under uniformly-distributed loading discussed in the preceding section is checked for determining whether the arrangement and amount of transverse reinforcement provided is sufficient for preventing any non-flexural type of failure occurring before the beam's flexural capacity is exhausted under two-point loading.

(a) *Load-carrying capacity*

When the beam attains its flexural capacity M_f, the shear force developing with the shear spans is $V_f = M_f/a_v = 445 \times 10^6/2,000 = 222500$ N. Thus, load-carrying capacity is $P_f = 2 V_f = 445000$ N.

(b) *Model*

For $a_v = 2,000$ mm, $a_v/d = 2,000/500 = 4 > 2.5$. Thus, the beam exhibits type I or type II behaviour and, hence, its model [shown in Fig. 4.4 (bottom)] is similar to that for the case of uniformly-distributed loading shown in Fig. 4.3, but, unlike the latter which may suffer brittle failure in the region of location 1, the former may also suffer such failure in the region of location 2.

(c) *Transverse reinforcement*

From expression 3.9, the 'shear' force that can be sustained without the need of transverse reinforcement in the region of location 1 is $V_{II,1} = 0.5 \times 2.37 \times 250 \times 500 = 148\ 125$ N < 222500 N. Thus, there is a need for transverse reinforcement capable of sustaining the whole transverse tensile force of 222500 N developing in this region, i.e. $A_{sv,III} = 222,500/500 = 445$ mm^2. Such reinforcement is placed within a length $2d = 2 \times 500 = 1,000$ mm, symmetrically with respect to location 1, the latter being at a distance equal to 2.5 $d = 1,250$ mm form the support closest to it. This amount of reinforcement is equivalent to five two-legged stirrups of 8 mm diameter at a spacing of 225 mm. For practical purposes, the specified amount of transverse reinforcement comprises two-legged stirrups of 8 mm

Fig. 4.4 Geometric characteristics (*top*) and physical model (*bottom*) of simply-supported beam exhibiting type II under two-point loading

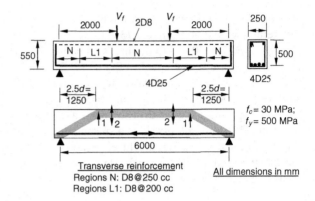

Transverse reinforcement
Regions N: D8@250 cc
Regions L1: D8@200 cc

All dimensions in mm

diameter at a spacing of 200 mm throughout the region [indicated as L1 in Fig. 4.4 (top)] of location 1.

From expression 3.13, the 'shear' force that can be sustained without the need of transverse reinforcement in the region of location 2 is $V_{II,2} = 982{,}000 \times [1-1/(1 + 5|2.37l/30)] = 278057$ N $> V_f = 222500$ N. Thus, there is no need for transverse reinforcement in the region of location 2 other than the nominal amount.

The transverse reinforcement details are shown in Fig. 4.4: Nominal reinforcement (such as that specified for the case discussed in the preceding section, i.e. two-legged 8 mm diameter stirrups at a spacing of 250 mm) is provided in the regions marked with 'N', whereas the reinforcement assessed as described in the preceding paragraph (i.e. two legged 8 mm diameter stirrups at a spacing of 200 mm) is placed within the regions of locations 1 marked with 'L1'.

4.5.3 Beam Under Two-Point Loading with $a_v/d = 3$

For the beam in Fig. 4.5, it is required to design transverse reinforcement capable of safeguarding against brittle types of failure.

(a) *Load-carrying capacity*

The beam's geometric characteristics, longitudinal reinforcement and material properties are those of the beams discussed in the preceding sections. Therefore, flexural capacity is $M_f = 445 \times 10^6$ and, hence, 'shear' capacity is $V_f = M_f/a_v = 445 \times 10^6/1{,}500 \approx 296667$ N. Thus, $P_f = 2 V_f \approx 593334$ N.

(b) *Model*

For $a_v = 1{,}500$ mm, $a_v/d = 1{,}500/500 = 3 > 2.5$. Thus, the beam exhibits type I or type II behaviour and, hence, its model [shown in Fig. 4.5 (bottom)] is similar to that shown in Fig. 4.4 (bottom).

Fig. 4.5 Geometric characteristics (*top*) and physical model (*bottom*) of simply-supported beam exhibiting type II behaviour under two-point loading

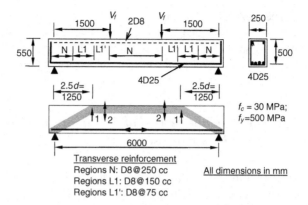

Transverse reinforcement
Regions N: D8@250 cc   All dimensions in mm
Regions L1: D8@150 cc
Regions L1': D8@75 cc

(c) *Transverse reinforcement*

From expression 3.9, $V_{II,1} = 0.5 \times 2.37 \times 250 \times 500 = 148125$ N < 296667 N. Thus, there is a need for transverse reinforcement capable of sustaining a transverse tensile force of 296667 N, i.e. $A_{sv,III} = 296{,}667/500 = 593.33$ mm². Such reinforcement, which should be placed within a length $d = 500$ mm on either side of location 1, is equivalent to six two-legged stirrups of 8 mm diameter (=6 x 1500.53 mm² = 603.18 mm²). Since the region of location 1 cannot extend beyond the load point, three of these stirrups are placed within a length of 500 mm to the left of location 1 at spacing of 150 mm, and three within a length of 250 mm to the right of location 1 (distance between location 1 and load point) at 75 mm spacing.

From expression 3.13, the 'shear' force that can be sustained without the need of transverse reinforcement in the region of location 2 is $V_{II,2} = 982{,}000 \times [1 - 1/(1 + 5|2.37|/30)] = 278057$ N $< V_f = 296667$ N. Thus, there is need for transverse reinforcement capable of sustaining the tensile stresses developing in the region of location 2 due to the loss of bond between concrete and the longitudinal reinforcement. The amount of reinforcement required for this purposed is calculated through the use of expressions (4.2) and (4.3). From expression (4.2), the transverse tensile stresses developing within the region of location 2 (extending from location 1 (i.e. the location of the joint of the horizontal and inclined elements of the 'frame') to the point load) are $\sigma_t = 30/[5 \times (982000/296667-1] \approx 2.6$ MPa, whereas from expression (4.3), the total tensile force developing within this region (with a length 250 mm) is $T_{II,2v} = 2.6 \times 250 \times 250/2 = 81250$ N in the vertical direction and $T_{II,2h} = 2.6 \times 9$ $4 \times 250/2 = 30550$ N horizontally. The amount of transverse reinforcement (within the length of 250 mm of the region of location 2) required to sustain $T_{II,2v}$ is $A_{sv,IIIv}$ $= 81250/500 = 162.5$ mm², i.e. two two-legged 8 mm diameter stirrups at a spacing of 125 mm, with the horizontal leg of the stirrups within the compressive zone being sufficient to sustain $T_{II,2h}$, since two one-legged 8 mm stirrups are capable of sustaining a force $F = 100.53 \times 500 = 50265$ N $> T_{II,2h} = 30550$ N. However, such reinforcement at location 2 is covered by that already specified for location1.

The transverse reinforcement details are shown in Fig. 4.5: Nominal reinforcement is provided in the regions marked with 'N', whereas the reinforcement assessed for location 1 is placed in the regions marked L1 and L1', with that
within L1' also covering the need for transverse reinforcement at location 2.

4.5.4 Beam Under Two-Point Loading Exhibiting Type III Behaviour

Figure 4.6 illustrates the geometric characteristics of a beam with type III behaviour (i.e. with a value of a_v/d (= 2,000/1,000 = 2) between 1 and 2.5), under two
point loading symmetrical about the mid-span. In the following the proposed
method is used to calculate the amount of transverse reinforcement required to
safeguard against brittle types of failure.

Figure 4.7 depicts the internal actions that would develop at any cross-section
within the portion of the beam between the load-points, were the beam capable to
reach its flexural capacity. Since the material characteristics are those already used
in the previous design examples, $\sigma_a = 41.84$ MPa, $F_c = \sigma_a bx = 41.84 \times 350 \times x =
14,644x$ and $F_s = 2 \times 982000$ N = 1964000 N. The equilibrium condition $F_c = F_s$
yields a depth of the compressive zone $x = 1964,000/(41.84 \times 350) \approx 134$ mm, and
thus the lever arm of the couple of the longitudinal internal forces is $z = d-x/2 = 1,
000-134/2 = 933$ mm. The moment of this couple yields the beam's flexural capac-
ity $M_f = 1,964,000 \times 933 = 1,832.41 \times 10^6$ Nmm.

Failure of the beam may also result from the reduction of the depth of the compressive zone owing to the extension of the inclined crack which is deeper than the
flexural cracks. (The causes of this type of failure—occurring as a result of failure
of the horizontal element of the 'frame', in the region of the joint of the horizontal

Fig. 4.6 Geometric
characteristics (*top*) and
physical model (*bottom*)
of simply-supported beam
exhibiting type III behaviour
under two-point loading

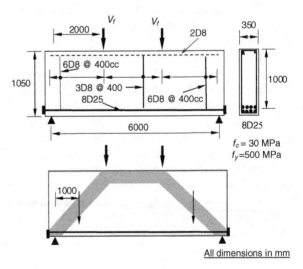

All dimensions in mm

Fig. 4.7 Internal actions developing at the portion of the model in Fig. 4.6 extending from the left-hand side support to the cross-section through the load closest to this support

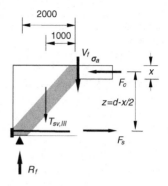

and inclined elements of the 'frame'—were described in Sect. 2.3). The reduction in depth causes a reduction in flexural capacity and the maximum bending moment M_{III} that can be sustained by the beam is assessed, as described in Sect. 3.3.3, by linear interpolation between two values of values of M_{III} corresponding to $a_v/d = 1$ ($M_{III} = M_f$) and $a_v/d = 2.5$ ($M_{III} = M_{fl}^{(2.5d)}$). Thus for $a_v/d = 2.5$, expression 3.9 yields $V_{II,1} = 0.5 \times 2.37 \times 350 \times 1{,}000 = 414750$ N from which it is easily obtained $M_{fl}^{(2.5d)} = 2.5dV_{II,1} = 2.5 \times 1{,}000 \times 414{,}750 = 1036.87 \times 10^6$ Nmm. As a result, for $a_v/d = 2$, expression (3.14) yields $M_{III} = 1{,}036.87 \times 10^6 + (1{,}832.41 \times 10^6 - 1{,}036.87 \times 10^6)(2.5 \times 1{,}000 - 2{,}000)/(1.5 \times 1{,}000) = 1{,}302.05 \times 10^6$ Nmm.

Comparing the above value with $M_f = 1{,}832.41 \times 10^6$ Nmm leads to the conclusion that the beam will fail before its flexural capacity is exhausted. As discussed in Sect. 4.3.2, this type of failure can be prevented by uniformly distributing transverse reinforcement in the form of stirrups within the whole length of the shear spans. The amount of such reinforcement required is assessed from expression (4.6) which yields $A_{sv,III} = 2(1{,}832.41 \times 10^6 - 1{,}302.05 \times 10^6)/(2{,}000 \times 500) = 1{,}062.72$ mm^2 required for this purpose, the latter being equivalent to twelve two-legged 8 mm diameter stirrups (with a total cross-sectional area of 1,206.37 mm^2) uniformly distributed within each of the beam's spans at a spacing of 200 mm; three additional two-legged 8 mm diameter stirrups at the same spacing are placed within the portion of the beam between the load points as nominal reinforcement [see Fig. 4.7 (top)].

4.5.5 Beam of Type IV Behaviour

Figure 4.8 illustrates the geometric characteristics of a beam with type IV behaviour (i.e. with a value of a_v/d (= 1,200/1,500 = 0.8) ≤ 1), under a single point load applied at mid span. In the following the proposed method is used to predict both the load-carrying capacity and the mode of failure of the beam.

As discussed in Sect. 3.3.4, the load-carrying capacity of the beam depends on the compressive strength of the weakest element of the 'frame' of the model

Fig. 4.8 Geometric
characteristics (*top*) and
physical model (*bottom*)
of simply-supported beam
exhibiting type IV behaviour
under point loading

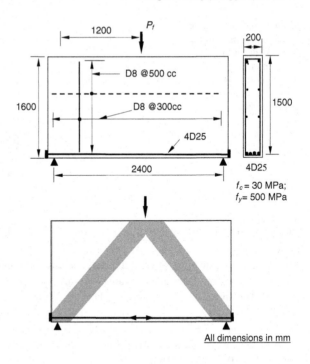

All dimensions in mm

Fig. 4.9 Internal actions
developing at the mid cross-
section of the model
in Fig. 4.8 (*bottom*)

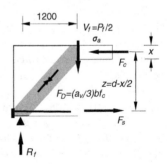

in Fig. 4.8 (bottom). Note that the single point loading precludes the formation
of a finite horizontal member of the 'frame' as explained in item (d) of Sect. 4.4.
From Fig. 4.9, the depth x of the cross-section of the 'horizontal' element (in this
instance, of course, the latter is made up entirely of the 'junction' in the 'frame')
is obtained from the equilibrium condition $F_c = F_s$, where $F_c = \sigma_a bx = 41.84 \times
200 \times x = 8{,}368x$ and $F_s = 982000$ N, since the material characteristics and
the longitudinal reinforcement are those already used in the previous design example; thus, $x = 982{,}000/(41.84 \times 200) \approx 117$ mm. With x known, the lever arm of
the couple of the longitudinal internal actions $F_c = F_s$ is $z = d - x/2 = 1{,}500 - 11$
$7/2 \approx 1{,}442$ mm. The condition $R_f a_v = F_s z$, where R_f is the reaction ($R_f = P_f/2$,
with P_f being the being the beam's load-carrying capacity), yields the value

$R_f = V_f = (z/a_v)F_s = (1,442/1,200) \times 982,000 \approx 1,180,037$ kN which corresponds to failure of the 'horizontal element of the 'frame' of the beam model.

On the other hand, the compressive strength of the inclined element of the 'frame' is exhausted when the axial force acting on it attains a value $F_D = (a_v/3)bf_c = (1,200 /3) \times 200 \times 30 = 2,400000$ N. The vertical component $P_D = F_D z/(z + a_v)^{1/2} = 2,400,000 \times 1,442/(1,442^2 + 1,500^2)^{1/2} = 1663276$ N is equal to the value of the reaction had the compressive strength of the inclined element being attained. Since the latter value is significantly larger than the value of the reaction corresponding to the beam's flexural capacity, the beam is predicted to exhibit a flexural mode of failure under a load $P_f = 2R_f = 2 \times 1180037 = 2360074$ N $\approx 2,360$ kN.

Due to its plate-like shape, the beam in Fig. 4.8 is sensitive to unintended out-of-plane actions which may lead to premature failure due to buckling. Such a type of failure is prevented through the provision of nominal reinforcement in the form of a cage comprising vertical two legged stirrups at a spacing of 300 mm and one layer of 8 mm horizontal bars at 500 mm spacing on either side face of the beam as indicated in Fig. 4.8.

Chapter 5
Design for Punching of Flat Slabs

5.1 Background

Punching, which may be suffered by two-dimensional (2D) reinforced-concrete (RC) structural elements (such as flat slabs, plates, footings, etc.) in regions under the action of concentrated load, is widely considered to be a 'shear' (non-flexural) type of failure. As a result, the methods adopted by current codes of practice (such as, for example, those adopted by ACI318 [1] and EC2 [2]) for designing against punching are essentially those applied to the 'shear' design of RC beam-like elements with modifications that allow for characteristics of structural behaviour particular to punching and to the geometry of the relevant structural elements. Furthermore, as for the case of 'shear' capacity, the assessment of punching capacity is invariably based on the use of semi-empirical formulae calibrated by regression analyses of published experimental data [1, 2].

Although such formulae appear to provide a realistic description of the effect of the element geometry and material properties on structural behaviour, their empirical nature precludes any insight into the fundamental causes of punching that the provision of transverse reinforcement is expected to prevent. Moreover, due to the large scatter of the experimental results, it is inevitable for the predicted values to exhibit significant deviations from individual test data. In fact, it is considered that instances of structural failures (such as, for example, the collapse of the Wolverhampton car park [3, 4]) may reflect the lack of a sound underlying theory rather than the inability of design formulae to provide a lower bound envelope to all experimental data.

On the other hand, as already discussed, the failure criteria presented in Chap. 3, are not only characterised by simplicity, but also provide a realistic description of the causes of non-flexural types of failure, as well as fit the experimental data published to date significantly closer than the formulae adopted by current codes. In the present chapter, the range of application of the above failure criteria is extended to cover the case of punching. In fact, it is shown that their use leads to design solutions which appear to be capable of safeguarding against punching.

M. D. Kotsovos, *Compressive Force-Path Method*, Engineering Materials,
DOI: 10.1007/978-3-319-00488-4_5, © Springer International Publishing Switzerland 2014

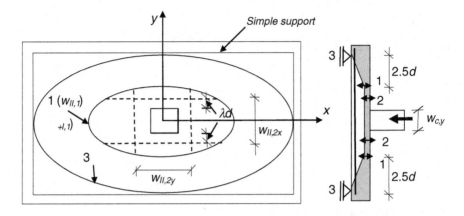

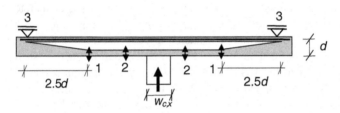

Fig. 5.1 Schematic representation of the physical state of a simply supported slab (considered to model the portion of the flat slab (in Fig. 5.2) encompassed by the section where the bending moment is zero (3)) indicating locations 1 and 2 of possible punching initiation

5.2 Criteria for Punching

The relevance to punching of both the causes of 'shear' failure of beam elements and the failure criteria proposed in Chap. 3 is discussed in what follows by reference to Fig. 5.1. The figure shows a rectangular simply-supported slab subjected to a patch load at its geometric centre, with the slab being considered to represent the middle portion of the larger slab shown in Fig. 5.2. From the latter figure, it can be seen that the above portion extends radially from the middle support (represented as a patch load in Fig. 5.1) to the geometric locus of the points where the bending moment diagrams drawn between this and the outer supports first intersect the slab. The geometric locus of the points of zero bending moment is also indicated in Fig. 5.1 where it is shown that its shape (marked with "3") deviates from the rectangular shape of the simple support as a result of the tendency of the corners to rise when the slab is under the action of the transverse load.

5.2.1 Punching Due to Bond Failure

It has been postulated that the causes of failure underlying the failure criteria discussed in Chap. 3 are also relevant to punching [5]. In fact, it has already been

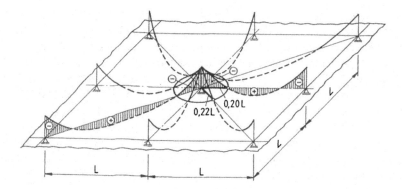

Fig. 5.2 Schematic representation of distribution of bending moment developing in a flat slab under transverse load

shown by analysis that punching of flat slabs is preceded by horizontal cracking of the compressive zone in the region of the slab's intersection with a supporting column (location 2 in Fig. 5.1), where large bending moments combine with large shear forces [6]; the causes for the occurrence of such cracking appear to be similar to those underlying the derivation of expression 3.13.

In order to use expression 3.13 for assessing the punching capacity of slabs, it has been further postulated that resistance to punching is predominantly provided by two slab strips intersecting at the column head and aligned in the (perpendicular to each other) directions of the flexural reinforcement [5]. These strips, which are considered to contain the flexural reinforcement (aligned in their direction) most likely to be the first (of the total amount of the slab flexural reinforcement in the direction considered) to yield, extend to a distance of λd on either side of the supporting column, as indicated in Fig. 5.1, i.e. the strip width will be

$$w_{II,2} = w_c + 2\lambda d \tag{5.1}$$

with

w_c being the column width along the axes of symmetry (x or y) of the column cross section;

d the effective depth of the slab; and

$\lambda,$ a parameter describing the effect of (1) concrete strength (f_c) and (2) the ratio (ρ) and the yield stress (f_y) of the flexural reinforcement

Guided by the results of a parametric study based on the use of a finite-element model that has been found capable of yielding realistic predictions of structural concrete behaviour [6], the parameter λ has been expressed in the form [5]

$$\lambda = \lambda_1 \lambda_2 = (2 - 100\,\rho f_y/500)\left[1 + 0.01\,(f_c - 60)\right] \tag{5.2}$$

with λ_1 and λ_2 being always not smaller than 1.

From expression 5.2, it can be seen that for the range of the types of concrete and steel currently used in practice and the values of reinforcement ratio commonly encountered in flat slabs, $\lambda = 1$, i.e. the strip width $w_{II,\,2}$ is independent of the material characteristics and amount of flexural reinforcement. On the other hand, a reduction of the tensile force sustained by the flexural reinforcement through the use of either a smaller reinforcement ratio ρ (i.e. $\rho < 1\,\%$), or steel with a smaller yield stress f_y (i.e. $f_y < 500$ MPa), appears to lead to an increase of the strip width $w_{II,2}$; the latter appears to also increase with the use of high-strength concrete, i.e. concrete with $f_c > 60$ MPa. The causes of the above effects are not as yet fully clarified.

Based on the above considerations, the shear force $V_{II,2}$ obtained from Eq. 3.13 corresponds to one of the four components (one for each side of the supporting column) of the punching capacity $P_{II,2}$, with F_c being the force developing within the compressive zone due to bending of the slab strip with a width $w_{II,2x}$ or $w_{II,2y}$ extending symmetrically about the column cross-section's axes of symmetry x and y, i.e.

$$P_{II,2} = \Sigma V_{II,2} = \Sigma F_{c^*} \left[1 - 1/\left(1 + 5|f_t|/f_c\right) \right] \tag{5.3}$$

5.2.2 Punching Initiation at the Location of Change in the CFP Direction

In contrast with the causes of failure underlying expression 3.13, those underlying expression 3.9 are not directly dependent of the flexural reinforcement, as they relate to the development of transverse tensile stresses in the region of the slab's locations where the compressive force path (CFP) changes direction (locations 1 in Fig. 5.1). As a result, resistance to the action of these stresses, in this case, is provided not by the slab strips discussed earlier but by the slab region extending to a distance d on either side of the above locations. The trace of these locations on the middle plane of the slab is encompassed by the geometric locus of the points where the slab bending moment becomes zero, with the two curves having a similar shape, since, as indicated in Fig. 5.1, the distance of the former from the latter is equal to $2.5d$. From the figure, it can be seen that, for the case of a rectangular slab, the trace of the locations where the CFP changes direction may be considered to have an elliptical shape, which may reduce to a circular one for the case of a circular or square slab.

In view of the above, if b is replaced with the perimeter $w_{II,1}$ of the trace (on the middle plane of the slab) of the locations at which the CFP changes direction, then $V_{II,1}$, assessed from expression 3.9, is considered to express the slab's resistance, $P_{II,1}$, to punching that may initiate in the region of the above locations due to the causes of failure underlying expression 3.9, i.e.

$$P_{II,1} = V_{II,1} = 0.5\ d\ w_{II,1} f_t \tag{5.4}$$

5.2.3 Punching Capacity

It appears from the above that punching may occur, as a result of the development of transverse tensile stresses within the compressive zone, either in the region of the column head (marked with "2" in Fig. 5.1) where large bending moments combine with large shear forces or in the region (marked with "1" in Fig. 5.1) where the path of the compressive forces (due to bending) emanating from the column-slab intersection change direction (at a distance equal to 2.5d from the location of the zero bending-moment) towards the opposite slab face. In the former case, resistance to punching, $P_{II,\,2}$, is assessed through the use of expression 5.3, whereas in the latter resistance to punching, $P_{II,\,1}$, results from expression 5.4, in both cases by implementing the modifications discussed in the preceding sections, with punching failure occurring when the acting punching force tends to exceed the smaller of $P_{II,\,1}$, and $P_{II,\,2}$.

It may be noted that, in contrast with current codes which do not provide any reasons to justify the particular perimeters around the column head where it is recommended to check for punching, the proposed criteria not only describe the causes of punching and the regions most likely to suffer it, but also specify the perimeter beyond which punching is unlikely to occur.

5.2.4 Verification of Proposed Criteria for Punching

The verification of the proposed criteria is based on a comparison of the predicted values of punching capacity with experimental values obtained from tests on simply-supported slabs under patch loading. As discussed in a preceding section, such slabs are considered to represent the portion of a flat slab, around a supporting column, which extends to the section where the bending moment becomes zero.

The ratio of the experimental to the calculated by the proposed method values of the punching load are given in a bar chart form in Figs. 5.3, 5.4, 5.5, 5.6, 5.7, 5.8, 5.9, 5.10, 5.11, 5.12, 5.13, 5.14, 5.15, 5.16, 5.17, and 5.18, with the figures also containing the designation of the specimens through which their design details may be obtained from the references cited in the figure captions (full details of the references indicated in Figs. 5.3, 5.4, 5.5, 5.6, 5.7, 5.8, 5.9, 5.10, 5.11, 5.12, 5.13, 5.14, 5.15, 5.16, 5.17, and 5.18 are provided in Ref. [7]). For purposes of comparison, the figures also include the ratios of the experimental values to those predicted by ACI [1] and EC2 [2]. In order to facilitate the comparison of the normalised values of punching capacity, the mean and standard deviation values of the information provided in each of the Figs. 5.3, 5.4, 5.5, 5.6, 5.7, 5.8, 5.9, 5.10, 5.11, 5.12, 5.13, 5.14, 5.15, 5.16, 5.17, and 5.18 are summarised in Table 5.1.

It should be noted that the design details of the slabs investigated cover a wide range of parameters, with f_c varying from a value as low as 11 MPa up to a value of over 115 MPa, f_y from about 300 MPa to around 750 MPa, ρ from 0.3 to 6.9 %, the slab span and effective depth d from 720 to 3,000 mm and 77 to 275 mm, respectively, and the column width/diameter (w_c) from 100 to 600 mm. The test values of punching capacity have been normalised with respect to the values assessed by the proposed criteria and their variation with each of the above parameters is shown in Figs. 5.19, 5.20, 5.21, 5.22, and 5.23. The figures also include the overall mean value of load-carrying capacity shown in Table 5.1.

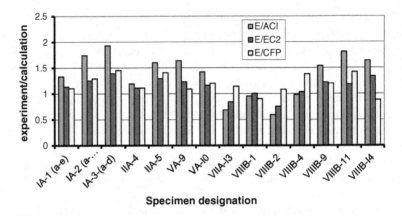

Fig. 5.3 Quadratic slabs tested by Elstner/Hognestad (1956) [7]: ratios of experimental (*E*) and calculated (through the proposed (CFP), ACI (ACI) and EC2 (EC) methods) values of punching load. (Average and standard deviation values of ratios for ACI, EC2 and proposed methods: 1.36, 1.13, 1.19 and 0.42, 0.18, 0.18, respectively)

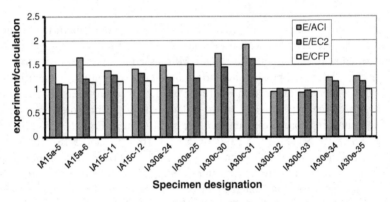

Fig. 5.4 *Circular* slabs tested by Kinnunen and Nylander (1960 [7]): ratios of experimental (*E*) and calculated (through the proposed (CFP), ACI (ACI) and EC2 (EC) methods) values of punching load. (Average and standard deviation values of ratios for ACI, EC2 and proposed methods: 1.41, 1.23, 1.06 and 0.3, 0.18, 0.09, respectively)

It is interesting to note that, in most cases, the proposed criteria predict that the occurrence of punching is linked with the causes of failure underlying expression 5.3, i.e. the loss of bond between concrete and the flexural reinforcement in the region of the column head where yielding of the reinforcement is likely to occur first. However, there are cases of punching in which failure initiates in the region of the locations where the compressive force due to bending changes direction due to the causes underlying expression 5.4, with the latter type of punching being more likely to occur for the case of low strength concrete ($f_c < 25$ MPa), particularly when it combines with values of ρ larger than 1 %.

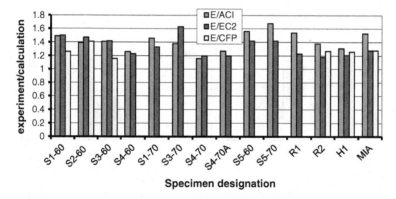

Fig. 5.5 Quadratic slabs tested by Moe (1961) [7]: ratios of experimental (*E*) and calculated (through the proposed (CFP), ACI (ACI) and EC2 (EC) methods) values of punching load. (Average and standard deviation values of ratios for ACI, EC2 and proposed methods: 1.41, 1.32, 1.16 and 0.13, 0.13, 0.18, respectively)

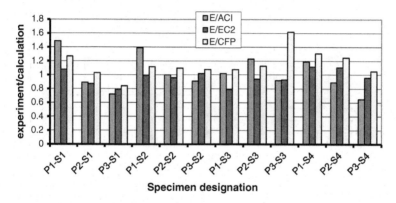

Fig. 5.6 Quadratic slabs tested by Manterola (1966) [7]: ratios of experimental (*E*) and calculated (through the proposed (CFP), ACI (ACI) and EC2 (EC) methods) values of punching load. (Average and standard deviation values of ratios for ACI, EC2 and proposed methods: 1.03, 0.95, 1.16 and 0.25, 0.11, 0.19, respectively)

From the figures, it can also be seen that for most slabs investigated the values predicted by the proposed criteria provide a closer fit to the experimental data than the values predicted by ACI, with the former values being similar to those predicted by EC2. In fact, Table 5.1 indicates that the overall mean value of those predicted by the proposed criteria is slightly better than its EC2 counterpart, with the latter being characterised by a slightly smaller standard deviation. On the other hand, the mean value of the ACI predictions exhibits not only a larger departure from the experimentally established values, but also the largest standard deviation.

Although other criteria (such as those, for example, described in Refs. [10–13]) have been shown to yield predictions similar to those of the criteria proposed in

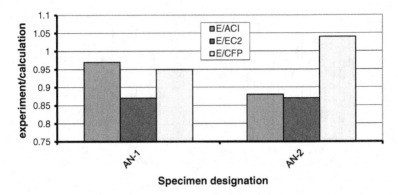

Fig. 5.7 Quadratic slabs tested by Corley/Hawkins (1968) [7]: ratios of experimental (E) and calculated (through the proposed (CFP), ACI (ACI) and EC2 (EC) methods) values of punching load. (Average and standard deviation values of ratios for ACI, EC2 and proposed methods: 1.0, 0.87, 0.93 and 0.06, 0, 0.06, respectively)

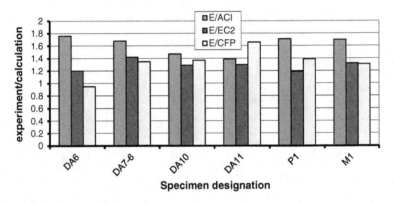

Fig. 5.8 *Circular* slabs tested by Ladner et al. (1970–1977) [7]: ratios of experimental (E) and calculated (through the proposed (CFP), ACI (ACI) and EC2 (EC) methods) values of punching load. (Average and standard deviation values of ratios for ACI, EC2 and proposed methods: 1.62, 1.29, 1.33 and 0.15, 0.09, 0.23, respectively)

the present work, it is considered that the advantages stemming from adopting the latter criteria relate to their ability to provide a better indication of the causes and location of failure, rather than a more accurate prediction of the punching load. In fact, as discussed in Sect. 5.3, the transverse reinforcement arrangement required to prevent punching failure due to the causes underlying the development of expressions 5.3 and 5.4 is significantly different than that resulting from the application of the code provisions and this improves structural behaviour.

It is interesting to note in Figs. 5.19, 5.20, 5.21, 5.22, and 5.23 that although the derivation of the proposed criteria is based on first principles without the need for calibration through the use of test data, the predicted behaviour provides a

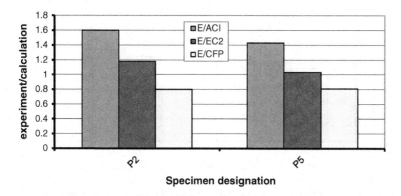

Fig. 5.9 *Circular* slabs P2 and P5 tested by ETH (1977/1979) [7]: ratios of experimental (*E*) and calculated (through the proposed (CFP), ACI (ACI) and EC2 (EC) methods) values of punching load. (Average and standard deviation values of ratios for ACI, EC2 and proposed methods: 1.52, 1.11, 0.81 and 0.12, 0.11, 0.01, respectively)

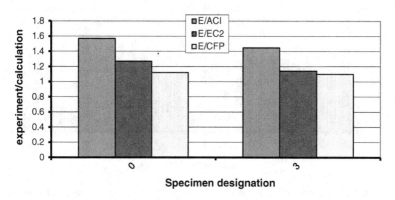

Fig. 5.10 *Circular* slabs tested by Schaefers (1978) [7]: ratios of experimental (*E*) and calculated (through the proposed (CFP), ACI (ACI) and EC2 (EC) methods) values of punching load. (Average and standard deviation values of ratios for ACI, EC2 and proposed methods: 1.51, 1.21, 1.11 and 0.09, 0.09, 0.01, respectively)

consistent description of the effect of the parameters involved in expressions 5.3 and 5.4. It is also interesting to note that the normalised test data exhibit a spread of the order of ± 0.5 about the mean value. This spread is similar to the deviation of the maximum and minimum values from the mean value of the tensile strength of concrete, f_t, which, as indicated in expressions 3.2a and 3.2b, is of the order of ± 0.45. Since the proposed criteria are directly linked to f_t, this compatibility between the spread of the predicted values of punching capacity and the spread of the experimentally-established values of the tensile strength of concrete is considered as an additional indication of the validity of the causes of punching expressed by the proposed criteria.

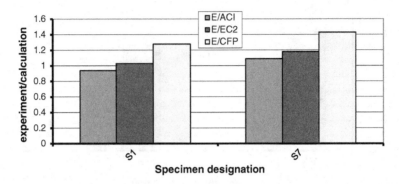

Fig. 5.11 Quadratic slabs tested by Swamy/Ali (1982) [7]: ratios of experimental (*E*) and calcu-lated (through the proposed (CFP), ACI (ACI) and EC2 (EC) methods) values of punching load. (Average and standard deviation values of ratios for ACI, EC2 and proposed methods: 1.02, 1.11, 1.36 and 0.11, 0.11, 0.11, respectively)

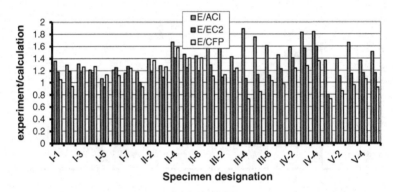

Fig. 5.12 Quadratic slabs tested by Regan (1986) [7]: ratios of experimental (*E*) and calcu-lated (through the proposed (CFP), ACI (ACI) and EC2 (EC) methods) values of punching load. (Average and standard deviation values of ratios for ACI, EC2 and proposed methods: 1.47, 1.19, 1.12 and 0.22, 0.17, 0.21, respectively)

5.3 Transverse Reinforcement for Punching

It has been proposed that the expressions used to assess the reinforcement required to prevent non flexural failure of beams can also be used to assess the transverse reinforcement required to prevent punching [14]. Such reinforcement is required when the design load is larger than the values resulting from expression 5.3 and 5.4. Reinforcement $A_{sv, II1}$ resulting from expression 4.1 is placed throughout the length $w_{II, 1}$ (see Sect. 5.2.2) and within a zone extending a distance d on either size of $w_{II, 1}$. On the other hand, reinforcement $A_{sv, II2}$ resulting from expressions 4.4 is distributed within the portion of the slab strip (with a width $w_{II, 2}$ resulting from expressions 5.1 and 5.2) extending from column-slab interface to the section

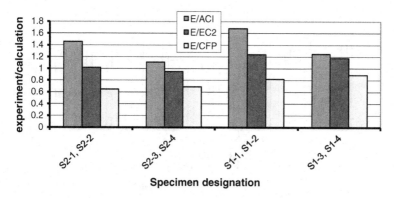

Fig. 5.13 *Circular* slabs tested by Tolf (1988) [7]: ratios of experimental (*E*) and calculated (through the proposed (CFP), ACI (ACI) and EC2 (EC) methods) values of punching load. (Average and standard deviation values of ratios for ACI, EC2 and proposed methods: 1.33, 1.11, 0.76 and 0.3, 0.14, 0.11, respectively)

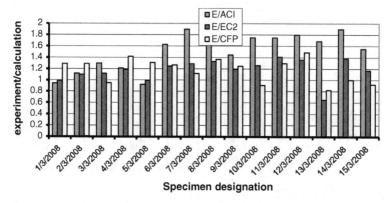

Fig. 5.14 *Circular* slabs tested by Ramdane (1993) [7]: ratios of experimental (*E*) and calculated (through the proposed (CFP), ACI (ACI) and EC2 (EC) methods) values of punching load. (Average and standard deviation values of ratios for ACI, EC2 and proposed methods: 1.51, 1.18, 1.18 and 0.33, 0.2, 0.21, respectively)

where the compressive force changes direction. Within the common area of the above regions only the larger of $A_{sv,\ II1}$ and $A_{ss,\ II2}$ is placed.

It is interesting to note that the method proposed above for assessing the transverse reinforcement required to prevent punching ignores the contribution of concrete to the slab's resistance to this type of failure. Such an approach adopts the reasoning that underlies the EC2 provisions for shear design, which has not as yet been extended to the code provisions for punching. It is possible for the proposed method, however, to allow for the contribution of concrete by subtracting its contribution ($V_{II,\ 1}$, in the region of locations 1, and $V_{II,\ 2}$, in the regions of locations 2, as obtained from expressions 5.4 and 5.3, respectively) from the corresponding values of V_f. The significance of this contribution forms part of a parametric study described later on.

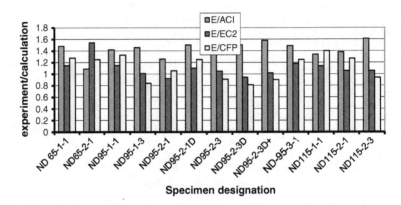

Fig. 5.15 Quadratic slabs tested by Tomaszewicz (1993) [7]: ratios of experimental (*E*) and calculated (through the proposed (CFP), ACI (ACI) and EC2 (EC) methods) values of punching load. (Average and standard deviation values of ratios for ACI, EC2 and proposed methods: 1.52, 1.07, 1.19 and 0.19, 0.08, 0.27, respectively)

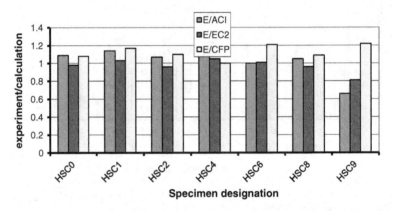

Fig. 5.16 *Circular* slabs tested by Hallgren (1996) [7]: ratios of experimental (*E*) and calculated (through the proposed (CFP), ACI (ACI) and EC2 (EC) methods) values of punching load. (Average and standard deviation values of ratios for ACI, EC2 and proposed methods: 1.1, 0.98, 1.1 and 0.26, 0.08, 0.1, respectively)

5.4 Verification of Design Method

5.4.1 Slabs Investigated

The verification of the proposed method is based on the comparative study of the results obtained from numerical tests on slabs designed in accordance with both the proposed method and the methods adopted by current codes such as the ACI318 and the EC2. The slabs are square in shape with a 2,000 mm side,

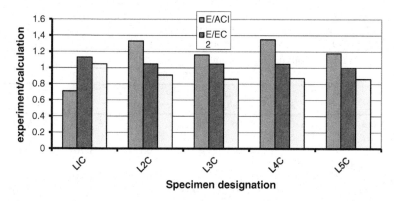

Fig. 5.17 *Rectangular* slabs tested by Olivera et al. (2004) [8]: ratios of experimental (*E*) and calculated (through the proposed (CFP), ACI (ACI) and EC2 (EC) methods) values of punching load. (Average and standard deviation values of ratios for ACI, EC2 and proposed methods: 1.19, 1.05, 0.91 and 0.08, 0.05, 0.08, respectively)

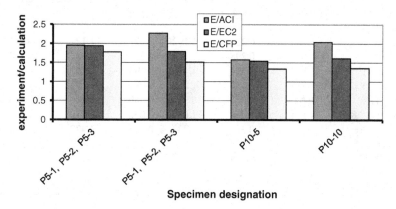

Fig. 5.18 Quadratic slabs tested by Papanikolaou et al. (2005) [9]: ratios of experimental (*E*) and calculated (through the proposed (CFP), ACI (ACI) and EC2 (EC) methods) values of punching load. (Average and standard deviation values of ratios for ACI, EC2 and proposed methods: 1.96, 1.72, 1.48 and 0.29, 0.18, 0.17, respectively)

a 200 mm depth, and a flexural reinforcement ratio (ρ) of 0.77 %. The uniaxial (cylinder) compressive strength of concrete and the yield stress of the longitudinal and shear reinforcement are $f_c = 30$ MPa, $f_y = 500$ MPa and $f_{yv} = 220$ MPa, respectively. The slab is considered to be monolithically connected to a column with a square cross-section of 400 mm side, and subjected to a monotonically increasing uniform displacement imposed on points arranged symmetrically about the slab's axes of symmetry so as to form a near circular curve with its centre coinciding with the geometric centre of the slab and its radius being approximately equal to 900 mm (see Figs. 5.24, 5.25, and 5.26).

The figures also provide an indication of the reinforcement arrangements resulting from the methods investigated, with Fig. 5.24 showing that the shear

Table 5.1 Summary of the mean and standard deviation values of the ratios of the experimental to predicted load-carrying capacities for the slabs in Figs. 5.3, 5.4, 5.5, 5.6, 5.7, 5.8, 5.9, 5.10, 5.11, 5.12, 5.13, 5.14, 5.15, 5.16, 5.17, and 5.18

Ref.	Load carrying capacity					
	EXP/ACI		EXP/EC2		EXP/CFP	
	Mean	Standard deviation	Mean	Standard deviation	Mean	Standard deviation
Kinnune and Nylander [7]	1.410	0296	1.228	0.181	1.061	0.090
Elstner and Hognestad [7]	1.325	0.419	1.129	0.182	1.204	0.161
Moe [7]	1.406	0.133	1.324	0.127	1.159	0.181
Monterola [7]	1.034	0.252	0.954	0.11	1.156	0.192
Corley and Hawkins [7]	1.000	0.064	0.870	0	0.925	0.064
Ladner et al. [7]	1.618	0.150	1.287	0.085	1.338	0.228
ETH [7]	1.515	0.120	1.110	0.106	0.81	0.007
Schaefers [7]	1.510	0.085	1.205	0.091	1.110	0.140
Swamy and Ali [7]	1.015	0.106	1.110	0.106	1.355	0.106
Regan [7]	1.466	0.221	1.192	0.17	1.12	0.211
Tolf [7]	1.325	0.298	1.100	0.135	0.763	0.112
Ramdane [7]	1.509	0.334	1.179	0.195	1.181	0.210
Tomaszewicz [7]	1.518	0.187	1.067	0.080	1.192	0.274
Hallgren [7]	1.100	0.260	0.982	0.080	1.100	0.098
Olivera et al. [8]	1.194	0.079	1.054	0.050	0.910	0.080
Papanikolaou et al. [9]	1.958	0.287	1.723	0.179	1.477	0.173
Overall mean and standard deviation values	1.389	0.314	1.162	0.201	1.133	0.209

reinforcement is distributed within the strips with a width $w_c + d$ intersecting at the column head and extending in parallel to the slab sides from the column-slab intersection to the section where the CFP changes direction. As discussed later, distributing the shear reinforcement within this strip width is found to produce the most effective design solution. In all cases the stirrup spacing is taken constant and equal to 100 mm, with the amount of stirrups being expressed in terms of the stirrup cross-sectional area rather than diameter for purposes of easier comparison. Depending on the method of design, the slabs are referred to as CFP, EC2 and ACI, whereas those without shear reinforcement as CON (control). It is interesting to note in the figures that the CFP method specifies a significantly larger amount of vertical reinforcement than that specified by current-code methods: moreover, the proposed method also specifies a layer of horizontally placed bars across the slab strips in the region of their compressive zone extending from the column-slab interface to the section where the compressive force changes direction.

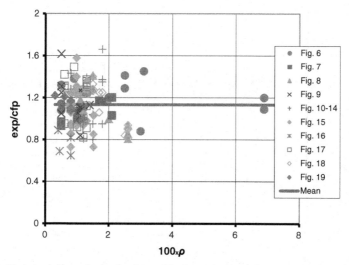

Fig. 5.19 Variation of experimental values of load-carrying capacity normalized with respect to their counterparts predicted by the proposed criteria with the flexural reinforcement ratio ρ

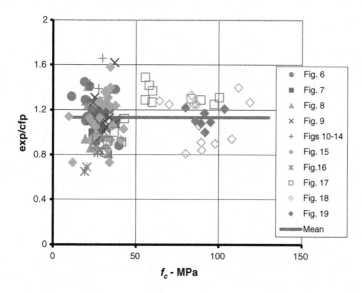

Fig. 5.20 Variation of experimental values of load-carrying capacity normalized with respect to their counterparts predicted by the proposed criteria with the concrete strength f_c

5.4.2 NLFEA Program Used for Verification

As discussed earlier, the behaviour of the flat slabs shown in Figs. 5.24, 5.25, and 5.26 is investigated by means of 3D NLFEA. The NLFEA program used for this purpose is fully described elsewhere [6, 15, 16] and, therefore, its presentation is beyond the scope of the present discussion.

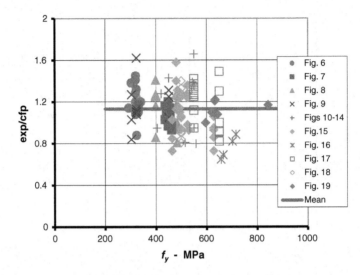

Fig. 5.21 Variation of experimental values of load-carrying capacity normalized with respect to their counterparts predicted by the proposed criteria with the yield stress of the flexural reinforcement f_c

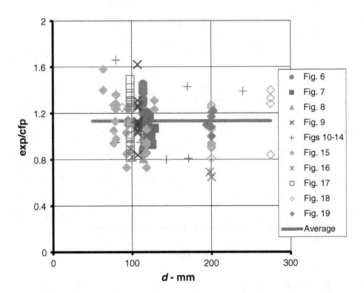

Fig. 5.22 Variation of experimental values of load-carrying capacity normalized with respect to their counterparts predicted by the proposed criteria with the effective depth d of the slab

Due to two-fold symmetry only one-quarter of the slabs is analysed, with this portion being discretized as shown in Fig. 5.27. Concrete is modelled by means of 27-node Lagrangian brick elements. Longitudinal and shear reinforcement is represented by 3-node line elements of appropriate cross-sectional areas possessing axial stiffness only.

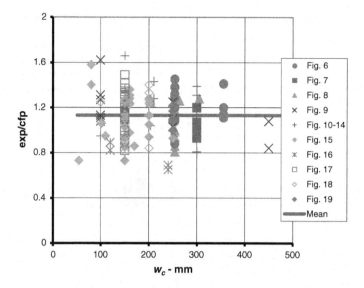

Fig. 5.23 Variation of experimental values of load-carrying capacity normalized with respect to their counterparts predicted by the proposed criteria with the column width/diameter w_c

The line elements used to model the steel reinforcement are not shown in the figure for clarity purposes; these elements are placed along consecutive nodes of the brick elements so as to maintain the amount and arrangement of various types of reinforcement per unit area equal to that of the slabs shown in Figs. 5.24, 5.25, and 5.26.

5.4.3 Results of Analysis and Discussion

The verification of the analysis package used has formed the subject of previous publications [6, 15, 16]; for the purposes of the present discussion, however, it is considered essential to provide additional evidence of the package's ability to yield realistic predictions of flat slab behaviour. Such evidence is provided in Fig. 5.28 which shows the load–displacement curves of two typical flat slabs (with and without shear reinforcement) with geometric characteristics and boundary conditions similar to those of the slab used as the basis for investigating the validity of the proposed design method. The figure also shows the experimentally-established values of load-carrying capacity, with the latter being also shown in Table 5.2 together with the slabs' design details; full design and test details are provided elsewhere [17]. From both figure and table, it can be seen that the predicted values of load-carrying capacity correlate closely with their experimental counterparts, with the former being smaller than the latter by less than 5 %.

Having established the package's ability to yield realistic predictions of slab behaviour, it is first used to establish the most effective spread (w_r) of the shear

Fig. 5.24 Design details of flat slab designed in accordance with proposed method

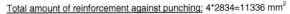

Total amount of reinforcement against punching: 4*2834=11336 mm²
——— Location of longitudinal reinforcement
⊗   Locations of induced displacement
•   Locations of shear reinforcement (2); 174mm²@100
•   Locations of shear reinforcement (1); 28mm²@100

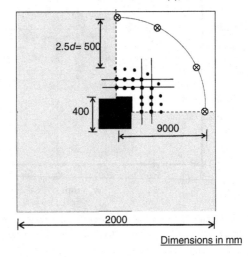

2.5d= 500
400
9000
2000
Dimensions in mm

Fig. 5.25 Design details of flat slab designed to ACI318

Total amount of reinforcement against punching: 4*1608=6432 mm²
⊗   Locations of induced displacement
•   Locations of shear reinforcement; 67mm²@100

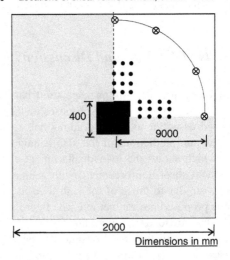

400
9000
2000
Dimensions in mm

reinforcement specified by the proposed method across the slab-strip width. Three cases of w_r are investigated: $w_r = w_c + 2d = 800$ mm, $w_r = w_c + d = 600$ mm, and $w_r = w_c = 400$ mm. The results obtained, expressed in the form of load–deflection

Fig. 5.26 Design details of
flat slab designed to EC2

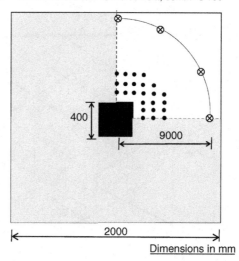

Total amount of reinforcement against punching:4*1272=5088 mm²
⊗   Locations of induced displacement
●   Locations of shear reinforcement; 53mm² @100

Dimensions in mm

relationships, are shown in Fig. 5.29. From the figure, it can be seen that distributing the shear reinforcement over a width $w_r = w_c + d = 600$ mm yields a small,
yet distinct, increase in load-carrying capacity. It is proposed, therefore, to distribute
the shear reinforcement specified by the proposed method across the slab strips to a
distance $(w_c + d)/2$ on either side of their axes of symmetry. This rule has been fol-
lowed thereafter when designing in accordance with the proposed method.

   An indication of the effect of the amount and arrangement of the vertical reinforcement specified by the proposed and the code methods investigated herein
is given in Fig. 5.30. The figure includes (a) the load–displacement relationship
predicted for the control slab (slab without shear reinforcement), (b) the value of
load-carrying capacity corresponding to flexural capacity assessed through the use
of the yield line theory and used as the basis for the design of the shear reinforcement, and (c) the value of the load corresponding to punching failure assessed
through the use of the proposed expressions. From the figure, it can be seen that
designing in accordance with the proposed method leads to a realistic prediction of
the load-carrying capacity, since the predicted value is smaller than the value corresponding to flexural capacity by less than 9 %. In fact, this margin may even be
smaller, as the latter value, as discussed above, has been assessed through the use
of the yield line theory which is known to produce values that tend to overestimate
load-carrying capacity. The value of load-carrying capacity predicted by analysis for the control slab appears also to correlate closely with the value assessed
through the use of expressions (5.3) and (5.4). On the other hand, designing the
'shear' reinforcement in accordance with the code adopted methods leads to

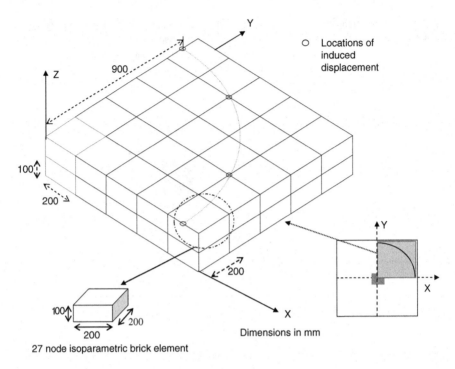

Fig. 5.27 Finite-element mesh of flat slab investigated

analytical predictions of the load-carrying capacity which are considerably smaller than the design value.

It could be argued that such a difference in behaviour may be attributed to the considerably larger amount of shear reinforcement specified by the proposed method, which is approximately twice as large as that specified by the code methods. And yet, Fig. 5.31 indicates that an increase of the amount of shear reinforcement designed in compliance to code provisions to the level specified by the proposed method, but without changing its arrangement, is essentially ineffective. It appears from the above, therefore, that the superior performance of the slab designed in accordance with the proposed method should be predominantly attributed to the specified arrangement, rather than amount, of the shear reinforcement.

The beneficial effect of the reinforcement arrangement specified by the proposed method is also demonstrated in Fig. 5.32. The figure shows the load–deflection curves of the slabs designed in compliance with the code specifications together with that of the slab designed in accordance with the proposed method modified (in the manner described in the preceding section) so as to allow for the contribution of concrete to the punching resistance of the slab. Although this modification results in a significant reduction of the amount of the shear reinforcement to a level comparable to the code specified amount, the figure indicates that it leads to a considerable improvement of the slab load-carrying capacity when

Table 5.2 Data for slabs and comparison between experimental and code predicted ultimate loads (P$_u$)

Slab ref.	Dimensions mm			Material properties mm		Main reinf	Stirrups	Pu, kN		Test/
	d	c	a	fc	fy	$\rho \times 10^{-3}$	mm^2	Test	Analysis	analysis
S2.1	200	250	2,400	24.2	657	8	–	603	583	1.03
S2.1s	195	250	2,400	24.9	501	8.2	2,513	894	848	1.05
S2.3/S2.4	200	250	2,400	25.4	668	3.4	–	489	490	1.00
S2.3s/S2.4s	198	250	2,400	24.7	671	3.4	1,256	552	545	1.01

Note d is the slab's effective depth; *c* is side of the square loaded area; *a* is the slab span; ρ is the flexural steel-reinforcement ratio

Fig. 5.28 Analytical relationships between load and displacement for two typical slabs with (denoted as 1) and without (denoted as 0) conventional shear reinforcement and experimentally-established values (indicated with the prefix E) of load-carrying capacity for the cases of flexural reinforcement ratios equal to 0.34 and 0.8 %

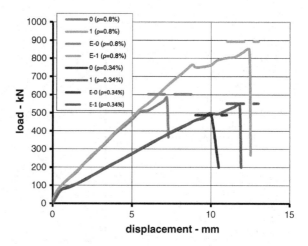

compared with the load-carrying capacity of the slabs designed to the code speci-fications. However, the results also indicate that the contribution of concrete is not as effective as the contribution of the additional reinforcement specified when the contribution of concrete is ignored.

An indication of the validity of the concepts underlying the proposed method may be obtained by investigating whether the causes of punching do indeed relate with the transverse tensile stresses developing within the com-pressive zone of the slab strips. This may be achieved by comparing the load–deflection curve obtained for a slab with the shear reinforcement within the slab strips extending throughout the slab depth with that of the same slab with the same shear reinforcement but this time extending to half the slab depth, the latter being slightly larger than the compressive zone depth. Such a com-parison is made in Fig. 5.33 which shows that placing shear reinforcement within the compressive zone only is sufficient for the slab to attain its design

Fig. 5.29 Analytical
relationships between load
and displacement for a
typical slab with the shear
reinforcement designed
in accordance with the
proposed method distributed
at various widths (800, 600,
and 400 mm) across the slab
strips specified by the design
method adopted

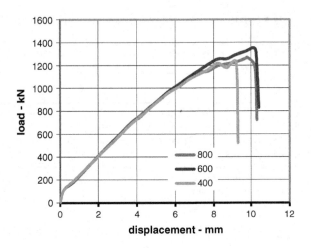

Fig. 5.30 Analytical load–
displacement curves for a
typical slab without (*curve*
CON) and with the shear
reinforcement designed in
accordance with the proposed
(*curve* CFP) and the ACI and
EC2 methods. Load-carrying
capacities assessed by the
proposed method for the
slab with and without shear
reinforcement are indicated
as CFP-D1 and CFP-D0,
respectively

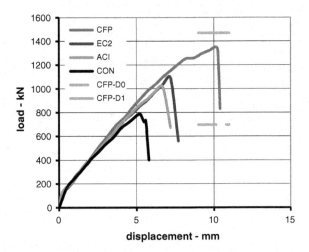

load-carrying capacity, thus confirming that the development of such tensile stresses is one of the underlying causes of punching.

An indication of the significance of the various types of shear reinforcement specified by the proposed method may be obtained through a comparison of the load–deflection curves shown in Fig. 5.34. From the figure, it can be seen that omitting either the horizontal compressive reinforcement (curve cfp-h) or the shear reinforcement in the region of the abrupt change in the path of the compressive force due to bending (curve cfp-v1) results in a small reduction of both the load-carrying capacity and the maximum deflection of the slab. The loss of load-carrying capacity becomes significant when reducing the shear reinforcement of the slab strips (curve cfp-v2) to a nominal amount, and this is a further indication of the significance of such reinforcement in preventing failure due to the development of transverse tensile stresses within the compressive zone. In fact, the slab

Fig. 5.31 Analytically established relationships load–displacement curves for a typical slab with the shear reinforcement arranged in accordance with the ACI and EC2 methods in the amounts specified by the proposed (*curves* ACIx1.83 and EC2x2.2) and the code methods (*curves* ACI and EC2)

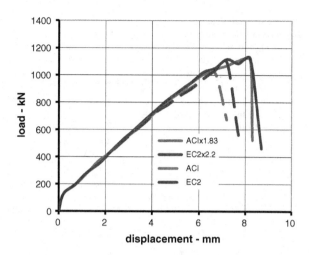

Fig. 5.32 Analytical relationships between load and displacement for a typical slab with the shear reinforcement designed in accordance with the ACI and EC2 and proposed (CFP) methods, as well as the proposed method modified so as to allow for the contribution of concrete to the slab resistance to punching (CFP-cc)

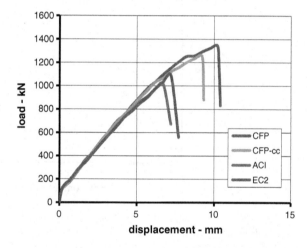

behaviour without such reinforcement is similar to that of the slabs designed in compliance with the code provisions, as indicated by comparing the relevant load–deflection curves in Figs. 5.30 and 5.34.

5.5 Concluding Remarks

The failure criteria and methods of assessment of the transverse reinforcement proposed in Chaps. 3 and 4, respectively, for beams are also shown to be valid for the case of punching. Comparing the values of the punching capacity of flat slabs assessed through the use of the proposed criteria and those recommended by ACI

Fig. 5.33 Analytical relationships between load and displacement for a typical slab with the shear reinforcement designed in accordance with the proposed method extending to either half (1/2) or the full (1) slab effective depth

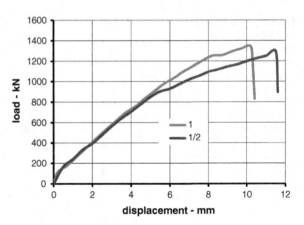

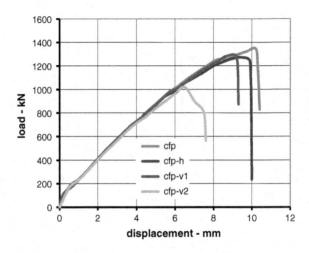

Fig. 5.34 Effect of the various types of shear reinforcement specified by the proposed method to the analytically established load–displacement relationships (*Notation* "cfp" indicates full compliance with design method; "cfp-h" slab without compression reinforcement; "cfp-v1" without the vertical reinforcement specified for the region of abrupt change in the direction of the CFP; "cfp-v2" with the amount of shear reinforcement specified for slab strips reduced to a nominal value)

and EC2 with a wide range of test data, shows that not only do the proposed criteria produce a closer fit to the test data, but also provide a more realistic description of the causes of punching initiation and an indication of its location. Also, the different, in both arrangement and quantity, reinforcement assessed through the use of the proposed method is found by finite element analysis, in contrast with the code methods, to achieve the design aim for load-carrying capacity. Moreover, the effect of the various types of transverse reinforcement specified by the proposed method on slab behaviour, as established by the numerical testing, provides evidence in support of the validity of the concepts underlying the proposed method.

References

1. American Concrete Institute (2002) Building code requirements for structural concrete (ACI 318-02) and commentary (ACI 318R-02)
2. EN 1992-1 (2004) Eurocode 2: design of concrete structures—part 1-1: general rules and rules for buildings
3. Kellermann JF (1997) Pipers row car park, Wolverhampton: results of the investigation. In: Proceedings of conference on "concrete car parks: design and maintenance issues" held at the Cavendish centre, British Cement Association, London, 29 Sep 1997
4. Shock collapse sparks lift slab fears and Safety experts urge car park review. New Civil Engineer, 27 Mar/3 Apr 1997, pp 3–4
5. Kotsovos GM, Kotsovos MD (2009) Flat slabs without shear reinforcement: criteria for punching. Struct Eng 87(23/24):32–38 1 Dec 2009
6. Kotsovos MD, Pavlovic MN (1995) Structural concrete: finite-element analysis for limit-state design. Thomas Telford, New York, p 550
7. Task Group 3.1/4.10, *Punching of structural concrete slabs*, Technical report, Bulletin 12, fib (CEB-FIP), April 2001, 307 pp
8. Oliveira DRC, Regan PE, Melo GSSA (2004) Punching resistance of RC slabs with rectangular columns. Mag Concr Res 56(3):123–138
9. Papanikolaou KV, Tegos IA, Kappos AJ (2005) Punching shear testing of reinforced concrete slabs, and design implications. Mag Concr Res 57(3):167–177
10. Long AE, Bond D (1967) Punching failure of reinforced concrete slabs. Proc ICE 37:139–135 May 1967
11. Long AE (1975) A two-phase approach to the prediction of the punching strength of slabs. ACI J 72(2):37–45
12. Rankin GIB, Long AE (1987) Predicting the punching strength of conventional slab-column specimens. Proc. ICE Part I 82:327–346 April 1987
13. Muttoni A (2008) Punching shear strength of reinforced concrete slabs without transverse reinforcement. ACI Struct J 105(4):440–450 July–August 2008
14. Kotsovos GM, Kotsovos MD (2010) A new design method for punching of RC slabs: verification by non-linear finite-element analysis. Struct Eng 88(8):20–25
15. Cotsovos DM, Pavlovic MN (2005) Numerical investigation of RC walls subjected to cyclic loading. Comput Concr 2:215–238
16. Cotsovos DM, Pavlovic MN (2006) Simplified FE model for RC structures under earthquakes. Proc Inst Civ Eng Struct Build 159(SB2):87–102 April 2006
17. Kinnunen S, Nylander H, Tolf P (2001) Undersokningar rorande genemostansning vid Institutionen for Byggnadsstatik, KTH, Nordisk Betong (Stockholm), No. 3, 1978, pp. 25–27.(cited in Task Group 3.1/4.10, Punching of structural concrete slabs. Technical report, Bulletin 12, fib (CEB-FIP), April 2001, 307 pp.)

Chapter 6
Design of Structures Comprising Beam-Like Elements

6.1 Introduction

In this chapter, the method proposed in Chap. 4 for the design of simply supported beams is shown to be applicable, not only to structural-concrete members other than simply-supported beams, such as, for example, cantilevers, fixed-end beams and connections, but also to any form of structural configuration comprising beam-like elements.

6.2 Structural Members Other than Simply-Supported Beams

6.2.1 Physical Models

Examples of the use of the physical model of a simply-supported beam shown in Fig. 3.1 for modelling other types of structural concrete members are illustrated in Figs. 6.1, 6.2, 6.3, and 6.4.

Cantilevers. Figure 6.1 (bottom) shows that a cantilever, subjected to a point load near its free end, may be designed as a simply-supported beam under a single point load applied at its mid cross-section [see Fig. 6.1 (top)]. This is because the boundary conditions at the fixed end of the cantilever are similar to the conditions developing at the mid-span cross section of the simply-supported beam [see Fig. 6.1 (middle)]. Similarly, a structural concrete wall under horizontal loading can also be designed through the use of the proposed methodology, since the wall may be visualised as a vertically oriented cantilever beam. In fact, the application of the proposed methodology to the design of structural concrete walls has been found to yield safe and efficient design solutions, in spite of the considerably smaller amount of transverse reinforcement required in comparison with that specified by current codes [1].

M. D. Kotsovos, *Compressive Force-Path Method*, Engineering Materials,
DOI: 10.1007/978-3-319-00488-4_6, © Springer International Publishing Switzerland 2014

Fig. 6.1 Use of physical model of a simply supported beam for modelling a cantilever (*top* simply-supported beam; *middle* portion of the beam in top between support and mid-span cross section; *bottom* cantilever beam)

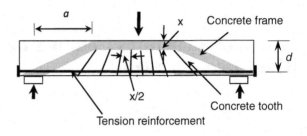

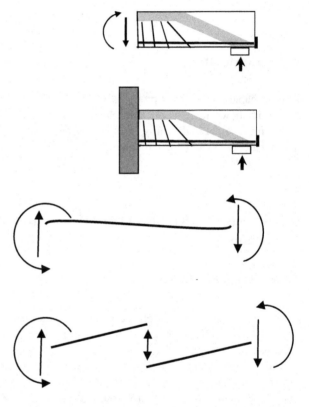

Fig. 6.2 Beam exhibiting a point of contraflexure in a structure and illustration of an internal tie needed at contraflexure in order to prevent separation of the two ends of the member

Fixed-end beams. Figure 6.2 depicts a reinforced concrete beam fixed at both ends, such as, for example, a beam coupling two structural walls. Such a beam, as for the case of the cantilever shown in Fig. 6.1, may also be designed through the use of the proposed methodology; the beam can be divided into two portions extending between the beam's fixed ends and the point of contraflexure, each of them essentially functioning as a cantilever. In this case, however, the design method must be complemented so as to allow for the design of the connection of the two 'cantilever' beams. Since the internal forces at the location of a point of contraflexure are equivalent to those developing at a hinged support, the connection between the two 'cantilevers' at this location may be viewed as an *internal support* that can be modelled as a transverse tie, through

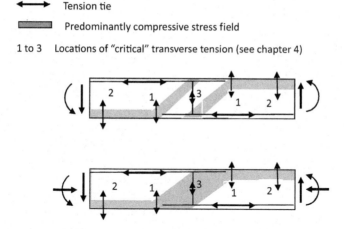

Fig. 6.3 Physical model of beam in Fig. 6.2 for the cases of $N = 0$ (*top*) and $N \neq 0$ (*bottom*)

which the continuity of the beam is safeguarded. The notion of the internal support (being modelled as a transverse tie) implies that concrete at the location of point of contraflexure is subjected to the action of transverse tension, rather than shear [2].

In the absence of axial force (i.e. $N = 0$), the model of the simply-supported beam in Fig. 6.1 (top) can be used for modelling the coupling beam in Fig. 6.2 as indicated in Fig. 6.3 (top). The latter figure shows that the physical model of the coupling beam comprises two cantilever beam models connected at the point of contraflexure, marked with "3" in the figure, by a transverse tie. Since concrete at location 3 is only subjected to transverse tension, its strength may be assessed through the use of expression (3.8) which is proposed in Chap. 3 for assessing the maximum value of the transverse tension sustained at location 1, i.e. the location of the link of the horizontal and inclined elements of the beam model (see Fig. 6.3). It should be noted that any load in excess of that that can be sustained by concrete at location 3 will cause the formation of a crack which, within the transverse tensile stress field prevailing in the region of this location, will extend in an unstable manner leading to immediate loss of load carrying capacity of the structural element [3, 4].

The effect of axial compression ($N > 0$) on the modelling of the beam in Fig. 6.2 is depicted in Fig. 6.3 (bottom), the latter essentially describing the mechanism of load transfer within the beam at its ultimate limit state. The figure shows that concrete at location 3 is subjected to transverse direct tension which develops within a predominantly compressive state of inclined stress, as opposed to the predominantly tensile transverse stress conditions developing in this region when $N = 0$ [see Fig. 6.3 (top)]. In both cases, the presence of transverse tension will inevitably lead to cracking when concrete strength is exhausted at location 3; but, unlike the latter case which, as already discussed, is linked with unstable crack extension leading to structural collapse, in the former case, crack extension will change direction so as to follow the path of the inclined compression and this will cause a stable, rather than unstable, crack extension process which delays structural collapse [3, 4].

Fig. 6.4 Schematic
representation of path of
inclined compression in
the region of the point of
contraflexure (location 3) for
the cases of a > 2.5d (*top*)
and a ≤ 2.5d (*bottom*)

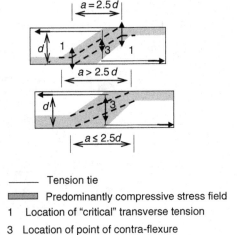

——— Tension tie

▬▬▬ Predominantly compressive stress field

1   Location of "critical" transverse tension

3   Location of point of contra-flexure

Beam-to-beam connection. Figure 6.3 (bottom) indicates the 'transfer' region through which the compressive force developing on account of bending combined with axial force is transferred from the right-hand side upper part of the beam to its left-hand side lower part. This transfer occurs along a path with a slope not smaller than about 1:2.5 (the latter being the slope of the inclined leg of the frame of the proposed model for simply-supported beams of type II behaviour discussed in Chap. 3) in a manner that will not violate the internal force conditions (i.e. the resultant of axial internal forces to be equal to N and bending moment equal to zero) developing at location 3 on account of the acting transverse and axial forces.

Graphical representations of the region of location 3 (internal support) for the cases of a length of this region $a > 2.5d$ and $a \leq 2.5d$ are indicated in Figs. 6.4 (*top*) and (*bottom*), respectively. From Fig. 6.4 (top), it can be seen that, for $a > 2.5d$, the slope of the inclined compression is 1:2.5, with locations 1 forming within the 'transfer' region on either side of location 3. Thus, in this case, the value of the transverse tension that can be sustained by concrete can be assessed as for type II behaviour through the use of expression (3.11).

On the other hand, when $a \leq 2.5d$, the slope of the inclined compression is $d{:}a \leq 1/2.5$. Such a slope is typical of type III behaviour and, therefore, the beam's load-carrying capacity can be assessed as discussed discussed in (Sect. 3.3.3).

6.2.2 Verification

It appears from the preceding section that the extension of the use of the model shown in Fig. 3.1 for the case of beam elements other than simply-supported requires the identification of the mechanism of the transfer of forces between successive beam-like elements. The mechanism proposed for this purpose in the preceding section is verified in the following through a comparison of the mechanism's predictions with experimental values. More specifically, experimental information obtained from the literature on the load-carrying capacity of structural members suffering failure in the region of a point of contraflexure is compared with its counterpart predicted by the proposed mechanism. Typical calculated and experimental values extracted from Ref. [5] are included in Tables 6.1 and 6.2. The tables include experimentally obtained values of load-carrying capacity expressed in the form of "shear" capacity (V_{EXP}) and corresponding axial compression (N_{EXP}), together with the values calculated from the proposed expressions (V_{PRE}), also corresponding to N_{EXP}. Examples of the application of the proposed method are provided in the following section for two typical cases included in Table 6.1. The values (V_{ACI}, V_{EC2}) calculated through the use of the formulae incorporated in ACI318 [6] and EC2 [7] are also included in the tables for purposes of comparison. The tables also contain the geometric characteristics and material properties required for assessing the specimen load-carrying capacity, with full details of the specimens tested being provided in the relevant publications cited in the tables.

Table 6.1 includes the results obtained from tests on simply-supported portal frames subjected, along their girder, to ten uniformly distributed vertical point loads combined with axial compression caused by two equal and opposite loads exerted at the lower end of the frame columns [8]. A schematic representation of the portal frame is provided in Fig. 6.5 which also includes geometric characteristics (complementing those in Table 6.1) essential for the assessment of the frame load-carrying capacity.

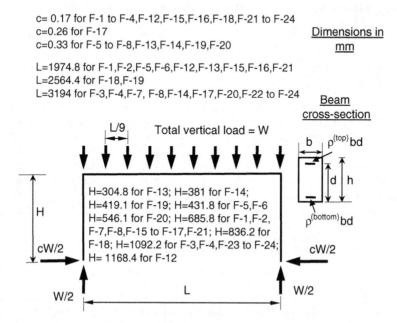

Fig. 6.5 Schematic representation of portal frames including geometric characteristics essential for the assessment of their load-carrying capacity given in Table 6.1

Table 6.1 Calculated and experimental values of shear capacity of frames suffering failure within the region of the point of contra-flexure [8]

Spec	ρ—% top/ bottom	f_y MPa top/ bottom	f_c MPa	N_{EXP} kN	V_{EXP} kN	V_{ACI} kN	V_{ACI}/V_{EXP}	V_{EC2} kN	V_{EC2}/V_{EXP}	V_{PRE} kN	V_{PRE}/V_{EXP}
Width (b) = 152.4 mm; height (h) = 304.8 mm; effective depth (d) = 254 mm; clear span length (l) = 1,828.8 mm											
Non flexural failure											
F-1	1.0/2.0	331/331	28.3	56	74	37	0.5	57	0.78	75	1.03
F-2	1.0/2.0	328/328	34.5	64	84	42	0.5	62	0.74	75	0,89
F-12	2.63/0.67	349/341	20.7	68	40	32	0.8	40	1	42	1.05
F-15	1.0/2.0	323/326	16.8	57	69	29	0.42	50	0.72	68	0.99
F-16	0.67/1.55	332/326	23.7	61	73	34	0.47	51	0.7	70	0.96
F-21	0.67/1.0	312/319	27.5	56	66	36	0.55	47	0.71	64	0.97
F-5	1.0/1.0	316/316	24.1	136	86	38	0.44	55	0.64	111	1.28
F-6	1.0/1.0	361/361	23.7	155	49	39	0.8	57	1.17	42	0.86
F-13	0.67/2	333/328	21.3	129	75	36	0.48	62	0.83	74	0.98
Mean values							0.55		0.81		1.00
Standard deviations							0.15		0.17		0.12
Width (b) = 152.4 mm; height (h) = 304.8 mm; effective depth (d) = 254 mm; clear span length (l) = 2,438.4 mm											
F-18	1.0/2.0	321/323	25.5	44	52	35	0.67	54	1.04	40	0.77
F-19	1.0/2.0	318/323	21	95	57	34	0.6	58	1.01	34	0.73
Mean values							0.64		1.02		0.75
Standard deviations							0.05		0.02		0.03
Width (b) = 152.4 mm; height (h) = 304.8 mm; effective depth (d) = 254 mm; clear span length (l) = 3,048 mm											
F-3	1.0/2.0	325/330	28.5	42	51	37	0.73	56	1.1	44	0.88
F-4	1.0/2.0	314/292	25.7	39	46	35	0.76	54	1.17		1.02
F-7	1.0/1.0	310/331	27.6	93	54	39	0.72	51	0.95	44	0.82
F-8	1.0/1.0	330/330	35.9	98	59	44	0.75	56	0.94	55	0.94
F-14	0.67/2.63	338/355	27.4	77	45	38	0.84	64	1.43	44	0.99
F-17	0.67/1.55	314/310	23	64	47	34	0.72	51	1.09	37	0.79
F-20	1.0/2.0	317/323	26.4	80	49	37	0.76	59	1.21	42	0.87
F-22	0.67/1.55	309/288	25.4	34	41	34	0.83	49	1.19	39	0.97
F-24	0.67/2.0	318/320	25.6	37	43	34	0.79	53	1.24	41	0.95
Mean value							0.77		1.15		0.91
Standard deviations							0.04		0.15		0.08
Mean value of all specimens							0.66		0.99		0.93
Standard deviations of all specimens							0.14		0.22		0.12
Flexural failure											
F-23	0.67/1.0	323/321	22.8	31	37					35	0.99

Table 6.1 indicates that the proposed expressions provide a close fit to the experimental values for types of behaviour II and III, and all values of N. In fact, the fit provided by the proposed expressions—developed, as already discussed in Chap. 3, from first principle without the need of calibration through the use of experimental data—is closer

Table 6.2 Calculated and experimental values of shear capacity of columns characterised by the formation of point of contra-flexure [9]

Spec	$\rho_v f_{yv}$ MPa	Width mm	f_c MPa	N_{EXP} kN	V_{EXP} kN	V_{ACI} kN	V_{ACI}/V_{EXP}	V_{EC2} kN	V_{EC2}/V_{EXP}	V_{PRE} kN	V_{PRE}/V_{EXP}
Height, $h = 365$ mm (375 for PC-1,PC-2,PC-3); effective depth, $d = 335$ mm (345 for PC-1, PC-2, PC-3); $\rho = 2.83$ %; $f_y = 487$ MPa; specimen length (l) = 1,310 mm											
Non-flexural failure											
PC-1	0.8	489	62.2	0	437	356	0.81	435	0.99	623	1.42
PC-2	0.8	489	62.2	3,452	863	657	0.76	750	0.87	715	0.83
PC-3	0.68	406	62.8	3,382	845	576	0.68	609	0.72	596	0.71
PC-4	0.72	375	83.1	0	401	212	0.53	338	0.84	478	1.19
PC-5	0.72	375	86.9	5,432	679	844	1.24	669	0.99	682	1.0
PC-6	0.72	375	86.9	2,657	668	559	0.84	669	1	536	0.8
PC-7	0.72	375	39.9	0	387	222	0.57	284	0.73	371	0.96
PC-8	0.72	375	42.4	1,986	497	369	0.74	448	0.9	438	0.88
PC-10	0.72	625	42.9	2,904	726	589	0.81	751	1.03	719	0.99
PC-11	0.72	625	44.9	4,130	754	690	0.91	769	1.02	813	1.08
PC-12	0.72	375	60	0	490	252	0.51	312	0.64	440	0.9
PC-13	0.72	375	60	2,720	680	484	0.71	539	0.79	524	0.78
PC-18	0.43	625	29.5	3,328	832	479	0.58	628	0.75	592	0.71
PC-20	0	625	47.3	4287	715	565	0.79	639	0.89	530	0.74
PC-21	0	625	47.3	3,071	767	473	0.62	639	0.83	447	0.58
PC-22	0.72	375	53.2	0	443	242	0.55	304	0.69	421	0.95
PC-23	0.72	375	53.2	1,808	603	388	0.64	504	0.84	464	0.77
PC-24	0.72	375	53.2	793	528	306	0.58	413	0.78	435	0.82
Mean values							**0.72**		**0.85**		**0.9**
Standard deviations							**0.18**		**0.12**		**0.2**
Flexural failure											
PC-9	0.72	375	42.9	4,130	510					643	099
PC-14	0.72	375	64.5	5,486	686					637	0.93
PC-15	0.72	375	56.6	7,157	358					329	0.92
PC-16	2.9	375	56.6	3,780	945					644	0.68
PC-17	2.9	375	56.6	5,654	707					546	0.77
PC-19	0.43	625	29.5	6,006	751					665	0.89
Mean value											**0.86**
Standard deviations											**0.12**
Mean value of all specimens											**0.89**
Standard deviations of all specimens											**0.18**

than that of the code formulae, which are empirical in nature and have been derived by regression analysis of experimental data, those of Tables 6.1 and 6.2 included.

The results shown in Table 6.2 are obtained from tests on beam-like elements subjected to the combined action of bending moment, shear force and axial compression at both ends [9]. Unlike the girder of the portal frame in Fig. 6.5, the specimens included web reinforcement; however, the quantity of such reinforcement 1s significantly smaller than the code specified amount for safeguarding a flexural mode of failure. Moreover, the values of axial compression exerted on the specimens are significantly larger than those in Table 6.1 and cover the whole range of practical values. The comparison between calculated and experimental values shows that the proposed expressions produce conservative predictions of load-carrying capacity with the smallest deviation from the experimentally obtained values for the whole range of practical values of axial compression. Moreover, it is interesting to note in the table that this deviation is similar to that characterising the deviation of the predicted values of flexural capacity from those established by experiment.

6.2.3 Typical Design Examples

Specimen PC-7 [9]. From Table 6.2, the geometric characteristics and material properties required for the calculation of the specimen's "shear" capacity are as follows: $b = 375$ mm, $d = 335$ mm, $h = 365$ mm, $l = 1{,}310$ mm, $f_c = 39.9$ MPa, $\rho_v f_{yv} = 0.72$ MPa.

The physical model of the specimen according to the CFP theory is shown in Fig. 6.6. Failure occurs at location 3 where the transverse tension $T_{(3)}$ is partly sustained by concrete (T_c) and partly by the transverse reinforcement (T_s) distributed over a length $2d = 670$ mm symmetrically extending about the location of the point of contraflexure (see Sect. 6.2.1). T_c is obtained from expression 3.8, with $f_t = 3.03$ MPa being obtained from expression 3.2a. Hence, $T_c = 0.5 \times 3.03 \times 375 \times 335 \times 10^{-3} = 191$ kN, whereas $T_s = 2\rho_v f_{yv} bd = 2 \times 0.72 \times 375 \times 335 \times 10^{-3} = 181$ kN, with T_s being obtained from expression 4.1. Therefore, $T_{(3)} = T_c + T_s = 191 + 181 = 372$ kN.

Since $T_{(3)} = V_{calc}$ and $V_{exp} = 387$ kN, $\mathbf{V_{calc}/V_{exp} = 0.96}$.

It should be noted that each of the two portions of the specimen, on either side of the point of contra-flexure, are characterised by type III behaviour, since $a_v = 1{,}310/2 = 655 < 2.5d = 2.5 \times 335 = 837.5$ mm. The 'shear' capacity of these portions is calculated through the use of expression 3.14 and found equal to 600 kN, which is nearly as large as the value corresponding to flexural capacity.

Specimen PC-8 [9]. From Table 6.2, the geometric characteristics and material properties required for the calculation of the specimen's "shear" capacity are as follows: $b = 375$ mm, $d = 335$ mm, $h = 365$ *mm*, $l = 1{,}310$ mm, $f_c = 42.4$ MPa, $\rho_v f_{yv} = 0.72$ MPa, $\rho = 2.83$ %.

The physical model of the specimen according to the CFP theory is shown in Fig. 6.7. The figure indicates that the length of the region through which the compressive force is transferred from the upper to the lower compressive zone extends

Fig. 6.6 Physical model of specimen PC-7 (in Table 6.2) according to the CFP theory

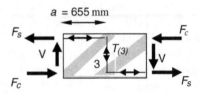

Fig. 6.7 Physical model of
specimen PC-8 (in Table 6.2)
according to the CFP theory

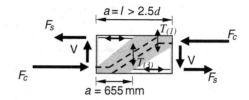

throughout the length of the element and, hence, it is larger than $2.5d$. In this case, as discussed in the preceding section, the slope of the inclined compression will be 1:2.5, and, therefore, location 1 will move within the transfer region. For such a slope of inclined compression, the transverse tension sustained by concrete (T_c) at location 1 can be calculated through the use of expression 3.11. Hence, replacing in expression (3.11) $k = (h - x_o)/(h - x_N) = 1.28$ (where $h = 365$ mm; $x_o = 47.54$ mm; $x_N = 117.27$ mm), $b = 375$ mm, and $d = 335$ mm results in $T_c = 257$ kN. The transverse tension sustained by the stirrups is $T_s = 2\rho_v f_{yv} bd = 2 \times 0.72 \times 375 \times 335 \times 10^{-3} = 181$ kN. Therefore, $T_{(1)} = T_c + T_s = 257 + 181 = 438$ kN.

Since $T_{(1)} = V_{calc}$ and $V_{exp} = 497$ kN, $V_{calc}/V_{exp} = 0.88$.

As for the case of specimen PC-7, the two portions of the specimen, on either side of the point of inflection, are characterised by type III behaviour, since $a_v = 1{,}310/2 = 655 < 2.5d = 2.5\cdot335 = 837.5$ mm and their 'shear' capacity is found equal to 600 kN, which is nearly as large as the value corresponding to flexural capacity.

6.3 Structural Configurations Comprising Beam-Like Elements

It becomes clear from Sect. 6.2 that a simply-supported beam may be viewed as any of the linear portions of a structural configuration such as, for example, the multi-storey frame depicted in Fig. 6.8, extending between consecutive points of zero bending moment (i.e. points corresponding to external simple supports or points of contraflexure). The figure shows that the frame essentially consists of four types of structural elements: columns extending between successive storeys (member A); beams spanning between successive columns (member B); beam-column joints (member C); and footings (member D). The case of footings is similar to the case of flat slabs which has already been discussed in Chap. 5, whereas the case of the columns is similar to the case of the fixed-end beams which has also been discussed in detail in Sect. 6.2.1 by reference to Fig. 6.3 (bottom). As regards members B and C, these have been isolated from Fig. 6.8 and depicted separately in Figs. 6.9 and 6.10, respectively; in the latter case, the beam-column joint sub-assemblage includes the portions of the adjacent beam and column elements extending to the points of contraflexure closest to the joint.

As for the case of columns, the case of a fixed-end beam, such as member B, has been discussed in Sect. 6.2.1, where it has been shown that a coupling beam can be divided into two portions extending between the beam's fixed ends and

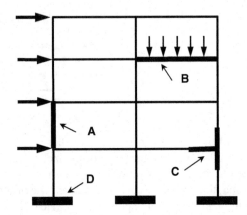

Fig. 6.8 Schematic representation of a multi-storey frame indicating two typical members: column (A), fixed-end beam (B), beam-column joint (C), and footing (D)

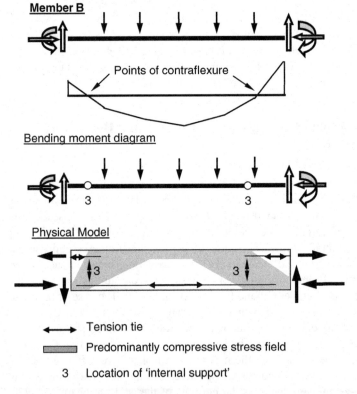

Fig. 6.9 Modelling of member B of the multi-storey frame in Fig. 6.8

the point of contraflexure, each of them essentially functioning as a cantilever. The link between the two 'cantilevers' at the point of contraflexure is viewed as an internal support that can modelled as a transverse tie. The fixed-end beam in

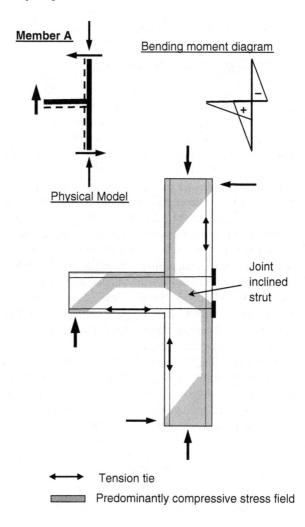

Fig. 6.10 Modelling of member C of multi-storey frame in Fig. 6.8

Fig. 6.9 (top) differs from that discussed in Sect. 6.2.1 in that it is subjected to a loading regime causing the formation of two points of contraflexure marked with "3" in Figs. 6.9 (middle) and 6.9 (bottom). Thus, the beam can be divided into three sub-elements: two cantilevers extending between each of the beam's ends and location 3 closest to it, and a simply-supported beam spanning the distance between locations 3 (see Fig. 6.9 (middle). Then, following the reasoning discussed in Sect. 6.2.1, the beam is modelled as indicated in Fig. 6.9 (bottom).

Similarly, for the beam-column joint sub-assemblage shown in Fig. 6.10, the portions of the beam and column elements connected to the joint are modelled and designed as discussed in Sect. 6.2; therefore, there will be no further discussion on the modelling and design of linear elements in this chapter. In what follows, attention will be focused on the modelling and design of a beam-column joint, which is the only member of the multi-storey frame in Fig. 6.8 that has not as yet been discussed.

6.4 Beam-Column Joints

6.4.1 Mechanisms of Load Transfer

The development of most models proposed to date for the design of beam-column joints is based on assumed mechanisms describing the manner in which the joint resists the action of the forces transferred to it by the adjacent beam-column elements. The most common of such mechanisms are the diagonal strut (indicated in Fig. 6.11a) and the truss (indicated in Fig. 6.11b) mechanisms which, at the ultimate limit state, are usually assumed to act concurrently [10–14]. The former mechanism is considered to resist the combined action of the normal and shear forces transferred to the joint through its interface with the compressive zone of the beam and column elements, whereas the later resists the action of the bond forces developing at the interface between concrete and the portion of the longitudinal beam and column reinforcement anchored within the joint.

The above mechanisms, which underlie the design model proposed by Park and Pauley [15] and adopted by the New Zealand code [16], have also been implicitly adopted by the current European codes EC2 [7] and EC8 [17] which specify means for safeguarding against shear failure of the joint occurring before the formation of a plastic hinge in the region of the beam adjacent to it. The code specifications include the permissible value of the shear force at the joint mid height, the anchorage length of the beam and column longitudinal steel bars and the amount and arrangement of the transverse reinforcement of the joint.

However, the application of current European code provisions for the design of earthquake-resistant joints has been found not only to lead to reinforcement

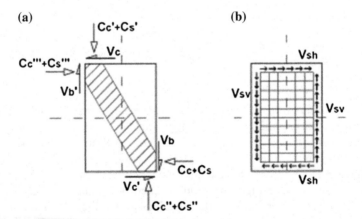

Fig. 6.11 Schematic representations of (**a**) diagonal strut and (**b**) truss mechanisms of joint resistance [in (**a**): C_c, C_c', C_c'', and C_c''' are the compressive forces in concrete; C_s, C_s', C_s'', and C_s''' are the compressive forces in steel; and V_b, V_b', V_c, V_c'' are the shear forces with subscripts b and c indicating the interfaces with beam and column, respectively. In (**b**): V_{sh} and V_{sv} the bond forces at the concrete-longitudinal steel of the beam-joint and column-joint interfaces, respectively]

congestion, and thus in difficulties in placing and compacting concrete, but also to design solutions not always satisfying the code performance requirements [18–22]. Attempts to improve design solutions have been primarily focussed on either the derivation of a more refined analytical description of the stress conditions linked with, or the implementation of modified versions of, the above mechanisms. In the following, only a limited number of such attempts are briefly discussed; they have been selected from the literature on the grounds that they differ from similar attempts in that they have produced design models which are expressed in a simplified form suitable for practical applications.

6.4.2 Design Models

A typical example of a design model resulting from the derivation of a more refined analytical description of the stress conditions related with the combined action of diagonal strut and truss mechanisms is the model proposed by Tsonos [14], which is, perhaps, the most suitable of such models for practical applications. A characteristic feature of this model is that it allows for the effect of the confinement provided by the transverse reinforcement on the compressive strength of concrete, the latter being directly linked with the load-carrying capacity of the diagonal strut; moreover, by expressing concrete strength as a function of the confinement provided by the transverse reinforcement, it allows the assessment of the transverse reinforcement required in order to adjust concrete strength so as to safeguard a predefined load-carrying capacity of the joint.

On the other hand, modified versions of the diagonal strut and truss mechanisms expressed in the form of strut and tie models, such as, for example, those indicated in Figs. 6.12 and 6.13, underlie the development of the analytical model developed by Hwang et al. [22] and the empirical design formulae proposed by Vollum and Newman [23], respectively. The strut and tie model shown in Fig. 6.12 has formed the basis for the development of an analytical approach which, through the use of a constitutive law for cracked reinforced concrete, satisfies both equilibrium and average strain compatibility conditions. However, the derived algorithm for assessing the joint shear capacity is complex and dependent on a large number of assumptions and empirical parameters for defining the geometric characteristics of the adopted strut and tie model, and this makes its application for the solution of common practical problems rather tedious. In fact, as pointed out by Vollum and Newman [23] and reiterated by Hegger et al. [24], the analysis and design of beam-column joints with strut and tie models are too complex owing to difficulties in determining node dimensions and the proportion of the shear force resisted by the stirrups. As a result, in their proposed simplified approach, the strut dimensions were established empirically from back analysis of selected test data and the validity of the resulting model (shown in Fig. 6.13) was demonstrated by analysing other test data; in fact, the model was found to predict joint shear strength more reliably than existing non-finite element methods and some finite-element techniques [23].

Fig. 6.12 Schematic representation of strut-and-tie model proposed by Hwang et al. [22]

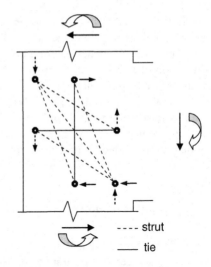

```
---- strut
____ tie
```

Fig. 6.13 Schematic representation of strut-and-tie model proposed by Vollum and Newman [23]

```
----    strut
____    tie
━━I     Beam &
        column bars
```

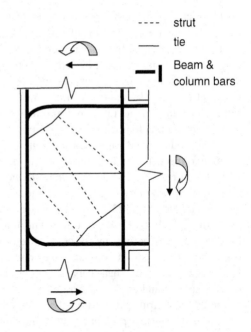

More recently, in an attempt to identify the causes of the observed behaviour of beam-column joints, it was shown by experiment that the bond between concrete and the portion of the members' flexural reinforcement anchored within the joint has an insignificant effect on both the crack pattern and the strength of the joint [25]. Moreover, it is realistic to expect that yielding of the flexural reinforcement of the beam and column members causes bond failure which extends deeply into the joint where the reinforcement is anchored and, as a result, the

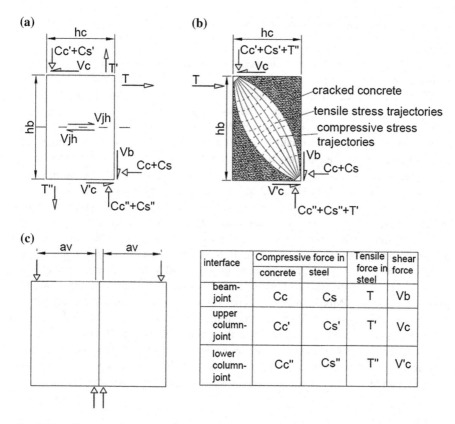

Fig. 6.14 a Forces acting at the interfaces of the joint with the beam and column elements and shear force V_{jh} at the horizontal cut at mid height of joint; **b** Compressive and tensile stresses trajectories assumed to develop within the joint at its ultimate limit state; and **c** Deep beam analogy

largest part of the tensile forces sustained by the reinforcement is directly transferred at the opposite side of the joint as indicated in Fig. 6.14. Such behaviour precludes the development of a truss mechanism and thus it is only through the diagonal strut mechanism that the joint resists the action of the forces transferred to it from the adjacent beam and column members [25] as suggested in the early 1980s [26].

The flow of the compressive stresses between the upper and lower diagonal ends of the joint is indicated by the compressive stress trajectories schematically represented in Fig. 6.14b. The figure also shows the tensile stress trajectories which intersect at right angles those of the compressive stresses. It is the flow of the compressive stresses that forms the diagonal strut which is confined between the regions of the cracked concrete surrounding the portion of the members' flexural reinforcement anchored into the joint (see also Fig. 6.14b); such cracking occurs under the action of the tensile forces which are transferred from steel to concrete and, as discussed earlier, eventually leads to loss of bond.

6.4.3 Application of CFP Model

In view of the above the load-bearing behaviour of the joint can be directly compared
with that of a deep beam, as suggested by Sarsam and Phipps [27] and Hegger et al.
[24]; thus, the strength of the diagonal strut may be assessed by visualizing the joint
in Fig. 6.14a as the shear span (a_v) of the deep beam under two-point loading shown
in Fig. 6.14c as proposed by Kotsovou and Mouzakis [25]. For such beams, which, as
discussed in Chap. 2, are essentially beams exhibiting type IV behaviour, the geometric
characteristics of the diagonal strut may be simplified by assuming that the latter has a
rectangular cross section (w_j $a_v/3$, where w_j is the beam width and a_v the shear span)
as recommended elsewhere [28], with the verification of the above recommendation
forming the subject of Ref. [29]. Adopting the aforementioned analogy for describing the load transfer within the joint (for which, as indicated in Fig. 6.15, $a_v = z_c$) the
design requirement for "rigid" beam-column joints is linked with the ability of the
diagonal concrete strut to safeguard the flow of compressive stresses (induced by
the forces indicated in Fig. 6.14b) between its upper and lower ends, without any significant cracking of the strut occurring before the formation of a plastic hinge at the
joint-beam interface. More specifically, it has been proposed that the flow of compressive stresses is safeguarded when the strut has sufficient strength to sustain the action
of the forces transferred to the joint from the beam and column elements, providing
sufficient reinforcement is placed to sustain the tensile forces developing along the tensile stress trajectories; this reinforcement will also contribute towards not only preventing the occurrence of excessive cracking, but, also, minimizing crack width [25].

6.4.4 Proposed Design Procedure

The above reasoning has been implemented in design through the following
design procedure [25], which is a refined version of the procedure described in
Sect. 3.3.4:

As indicated in Fig. 6.14a, the resultant of the horizontal forces transferred to
the upper side of the joint by the beam and column members is

$$F_{jh} = T - V_c = 1.2 A_{sl,b} f_y - V_c \qquad (6.1)$$

where $A_{sl,b}$ is the total area of the beam top longitudinal reinforcement, V_c is the
value of the shear force at the upper column-joint interface corresponding to the formation of a plastic hinge at the beam-joint interface, and '1.2' is a code specified
factor for the enhancement of the tensile force sustained by the beam longitudinal
steel reinforcement so as to minimize, if not eliminate, the deviation of the calculated code value of flexural capacity from its true counterpart. When flexural capacity is assessed through the method proposed in Chap. 3, the '1.2' factor is omitted.

The combined action of force F_{jh} with the compressive force (C_c) developing in
the compressive zone of the upper column on account of bending and transferred to

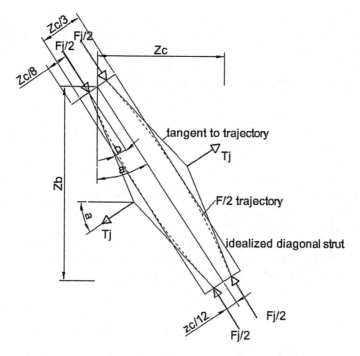

Fig. 6.15 Diagonal strut and acting stress resultants

the joint is balanced by the compressive force F_j developing along the diagonal strut (see Fig. 6.15); the latter is obtained by considering the equilibrium of the horizontal component of the forces acting at the upper end of the diagonal strut, i.e.,

$$F_j = F_{jh}/sina \qquad (6.2)$$

where α is the inclination of the diagonal strut shown in Fig. 6.15.

The value of F_j cannot be larger than the strength ($F_{Rj,max}$) of the diagonal strut under monotonic loading, i.e.,

$$F_j \leq F_{Rj,max} \qquad (6.3)$$

where, as proposed elsewhere (2,18), $F_{Rj,max}$ is obtained from expression

$$F_{Rj,max} = (z_c/3)w_jf_c = [(h_c - x_c)/3]\ w_jf_c \qquad (6.4)$$

where h_c is the column cross section height,
x_c the depth of the column compressive zone depth,
$z_c = h_c - x_c$, and
w_j the joint width

It should be noted, however, that when $x_c > (h_c - x_c)/3$, then $(h_c - x_c)/3$ in expression 6.4 is replaced with x_c.

However, due to the curved shape of the stress trajectories along which the compressive force developing within the diagonal strut is transferred from the upper left-hand side to the lower right-hand side of the joint, there is a tensile stress resultant developing across the diagonal strut at mid height of the joint. It has been suggested [25] that this tensile stress resultant may be obtained from

$$T_j = F_j \tan b \tag{6.5}$$

where b is the angle between the joint diagonal and the tangents at the ends of any of the two symmetrical (with respect to the joint diagonal) trajectories of the two compressive stress resultants, $F_j/2$, the latter acting at a distance equal to $z_c/12$ from joint diagonal on either side of it (see Fig. 6.15). Angle b is easily assessed by assuming that the tangents at the ends of the compressive-stress resultants' trajectories intersect at a distance equal to $z_c/8$ from joint diagonal (see Fig. 6.15).

The development of T_j is likely to cause cracking and, therefore, reinforcement should be placed in order to sustain it when the tensile strength of concrete is exhausted. The amount of reinforcement, which may be either inclined ($A_{sj,incl}$) or horizontal ($A_{sj,h}$), required to sustain T_j is obtained from

$$A_{sj,incl} = T_j / f_y \tag{6.6}$$

$$A_{sj,h} = T_j / (f_y \cos a) \tag{6.7}$$

As discussed earlier, the above reinforcement is also required because it contributes towards preventing the occurrence of excessive cracking, as well as minimizes crack width.

The force $F_{Rj,s}$ that can be sustained along the diagonal strut when the transverse reinforcement is at yield is obtained from expression

$$F_{Rj.s} = T_j / \tan b \tag{6.8}$$

Therefore, the maximum design force that can be sustained by the diagonal strut is

$$F_{jd} \leq min \left(F_{Rj,max}, F_{Rj.s} \right) \tag{6.9}$$

6.4.5 Verification of Design Method for Beam-Column Joints

Ideally, a beam-column joint should behave as a rigid body; as this, in reality, is not possible, the performance requirements of current codes are considered to be satisfied when the formation of a 'plastic hinge' in the beams adjacent to a joint occurs before the latter suffers significant cracking, i.e. cracking capable of having a statistically measurable effect on structural behaviour. An investigation of the ability of the method proposed in the preceding section to satisfy the above performance requirement has been based on the comparison of the structural behaviour predicted through the use of expressions 6.1–6.7 with its experimental counterpart reported in the literature for a wide range of beam-column joint sub-assemblages under cyclic loading [30]. The comparison has also included predictions of code adopted formulae, such as those adopted by ACI318 [6] and EC2 [7]–EC8 [17], as well as the predictions of two widely referred to models of the behaviour of beam-column joints (those proposed by Tsonos [14] and Vollum and Newman [23]), which, not only are they typical of those developed on the basis of the combined diagonal strut and truss mechanisms or strut-and tie models, but also expressed in a simple form suitable for practical structural design.

An indication of the results of the above comparative study is provided in Tables 6.3, 6.4, and 6.5 extracted from Ref. [30]. The tables include the predicted modes of failure and values of the joint capacity (V_{Rj}) together with their experimental counterparts obtained from the literature. The modes of failures are classified as either *formation of a plastic hinge in the beam* (BF) or *excessive cracking of the joint* (JF). BF is considered to preclude JF, since the latter may only occur after a loss of load-carrying capacity (due flexural capacity degradation) well beyond that allowed by current codes; on the other hand JF is considered to occur before, or concurrently with, BF [25], with concurrent occurrence of BF and JF (denoted as JF-BF) being considered as JF. Full design characteristics of the specimens can be found in the relevant publications cited in the tables. It should also be noted that the specimens selected for the comparative study have been obtained from publications which provide experimental information clearly describing the specimen mode of failure. Moreover, specimens in which the anchorage details of the beam flexural bars sharply deviate from those adopted in current design practice have not been considered.

From Table 6.3, it can be seen that, for the 74 cases investigated, EC2 [7]–EC8 [17] produced 42 successful predictions of the specimens' mode of failure, ACI318 [6] 32, the Vollum and Newman model [23] 41, the Tsonos model [14] 52, and the proposed method 72. It should also be noted that for specimens I and I-A tested by Hanson and Conner [31], EC2 [7]–EC8 [17] was unable to produce any predictions, since the axial load applied to the specimens was unrealistically high and outside the range of application of the code formulae. On the other hand, for specimens 2T5 and 1T55, the expressions proposed in the preceding section appear to underestimate, rather than overestimate, joint capacity by an amount of the order of 10 %. It would appear, therefore, that, unlike the other formulae, those proposed in the preceding section yield safe design solutions for all specimens investigated.

It is also interesting to note in Table 6.3 that, with the exception of the amount of transverse reinforcement specified by ACI318 [6] for specimens J0, JS, JX10, and JX12, the formulae proposed in the preceding section specify an amount of such reinforcement which is considerably less than the amount specified by either of the two codes considered. In fact, the amount specified by the proposed expressions is usually less than half and, in certain cases, less than a quarter the code specified amount. It would appear, therefore, that the above expressions are capable of, not only safeguarding the code specified performance requirements, but, also, reducing considerably reinforcement congestion in the joint.

Table 6.4 contains information extracted from Table 6.3; this information includes only the formulae predictions and the experimental data for the specimens characterized by excessive cracking of the joint (JF). From the table, it can be seen that, of the 37 cases included in the Table, the formulae adopted by EC2 [7]–EC8 [17] and ACI318 [6]

Table 6.3 Comparison of predictions with published experimental information on beam-column sub-assemblages characterised by failure of the join

Specimen designation (total no of specimens: 51) | Horizontal joint strength—V_{jRh} (kN); code specified transverse reinforcement—$A_{sv,c}$ (mm²); existing transverse reinforcement—$A_{sv,exp}$ (mm²); horizontal component of maximum sustained force by incline strut—$F_{jmax,h}$ (kN); horizontal component of compressive force corresponding to tensile force sustained by existing transverse reinforcement—$F_{j,h}$ (kN); failure mode—FM (joint failure—JF; beam failure—BF; experimentally established joint strength—$V_{j,exp}$ (kN)

Transverse reinforcement normally in the form of stirrups; suffix "d" indicates diagonal form of transverse reinforcement

Specimen	EC2-EC8			ACI 318			Vollume and Newman		Tsonos		Kotsovou and Mouzakis				Experimental results		
	V_{jRh}	A_{svc}	FM	V_{jRh}	A_{svc}	FM	V_{jRh}	FM	V_{jRh}	FM	$F_{jmax,h}$	$F_{j,h}$	V_{jRh}	FM	$V_{j,exp}$	$A_{sv,exp}$	FM
Hanson and Conner [31]																	
I	–	1,276	–	1,362	1,688	JF	1,027	BF	855	JF	1,205	1,477	1,205	BF	972	1,267	BF
I-A	–	906	–	1,314	1,387	JF	857	JF	747	JF	1,128	947	947	BF	915	713	BF
II	1,395	1,995	JF	1,390	1,496	JF	1,111	BF	908	BF	1,071	1,640	1,071	BF	908	1,267	BF
Kurose [32]																	
S41	949	1,791	JF	688	676	JF	435	JF	299	JF	397	343	343	JF	408	382	JF
S42	949	1,791	JF	688	676	BF	547	BF	333	JF	397	685	397	JF	402	763	JF
U41L	1,029	1,791	JF	720	742	JF	450	JF	325	JF	439	343	343	JF	413	382	JF
Eshani and Wight [20]																	
1B	945	956	JF	806	1,137	JF	617	JF	438	JF	475	1,182	475	JF	700	865	JF
2B	946	935	JF	822	1,083	JF	668	JF	444	JF	538	1,149	538	JF	707	865	JF
3B	1,102	953	BF	889	1,557	JF	681	BF	548	JF	600	1,763	600	JF	671	1,298	JF
4B	1,194	962	BF	929	1,555	JF	755	BF	574	JF	684	1,700	684	BF	681	1,298	BF
5B	929	1,762	JF	892	590	BF	736	JF	486	JF	561	1,164	561	JF	831	865	JF
6B	1,549	963	BF	1,141	1,100	JF	941	BF	682	BF	861	1,094	861	BF	608	865	BF
Eshani and Alameddine [33]																	
LH8	2,134	2,274	BF	1,288	2,248	BF	1,207	BF	1,011	BF	1,320	2,854	1,320	BF	920	2,280	BF
LH11	2,612	2,316	JF	1,288	2,248	BF	1,388	BF	1,392	BF	1,947	2,636	1,947	BF	898	2,280	BF
LH14	3,007	2,357	JF	1,288	2,324	JF	1,565	BF	1,755	BF	2,676	2,451	2,451	BF	907	2,280	BF
HL8	1,962	2,732	JF	1,288	2,248	JF	1,207	BF	926	JF	1,166	2,029	1,166	JF	1,185	1,520	JF
HH8	1,962	2,732	JF	1,288	2,248	BF	1,207	BF	1,011	JF	1,166	3,045	1,166	BF	1,172	2,280	JF

(continued)

Table 6.3 (continued)

Specimen designation (total no of specimens: 51): Horizontal joint strength—V_{jRh} (kN); code specified transverse reinforcement—$A_{sv,c}$ (mm²); existing transverse reinforcement—A_{sv} (mm²); horizontal component of maximum sustained force by incline strut—$F_{jmax,h}$ (kN); horizontal component of compressive force corresponding to tensile force sustained by existing transverse reinforcement—$F_{j,h}$ (kN); failure mode—FM (joint failure—JF; beam failure—BF); experimentally established joint strength—$V_{j,exp}$ (kN)

Transverse reinforcement normally in the form of stirrups; suffix "d" indicates diagonal form of transverse reinforcement

Specimen	EC2–EC8			ACI 318			Vollume and Newman		Tsonos		Kotsovou and Mouzakis				Experimental results		
	V_{jRh}	$A_{sv,c}$	FM	V_{jRh}	$A_{sv,c}$	FM	V_{jRh}	FM	V_{jRh}	FM	$F_{jmax,h}$	$F_{j,h}$	V_{jRh}	FM	$V_{j,exp}$	$A_{sv,exp}$	FM
Tsonos et al. [21]																	
S1	515	355	JF	404	506	JF	300	BF	200	BF	302	402	302	BF	144	302	BF
X1	375	349	BF	339	356	BF	253	BF	152	BF	185	1,040	185	BF	136	302s 308d	BF
S2	379	284	BF	339	356	BF	253	BF	152	JF	210	431	210	BF	157	302	BF
X2	352	283	BF	325	329	BF	243	JF	144	JF	187	1,037	187	BF	154	302s 308d	BF
S6	464	703	JF	381	451	JF	285	JF	182	JF	225	437	225	JF	314	302	JF
S6'	407	692	JF	357	369	JF	267	BF	165	JF	165	497	165	JF	230	302	JF
X6	381	689	JF	345	396	BF	258	JF	157	JF	166	1,090	166	JF	298	302s 308d	JF
Chutarat and Aboutaha [34]																	
I	2,074	2,316	JF	1,350	668	BF	1,154	BF	976	JF	769	1,711	769	JF	986	2,027	JF
A	3,137	905	BF	1,849	904	BF	1,580	BF	1,161	BF	1,320	1,711	1,320	BF	370	2,027	BF
Hwang et al. [22]																	
0T0	4,091	2,058	JF	1,672	2,138	JF	1,408	BF	1,202	BF	2,532	–		JF	896	0	JF
3T44	4,085	2,058	JF	1,672	2,408	JF	2,056	BF	1,814	BF	3,061	3,019	3,019	BF	886	2,280	BF
3T3	3,910	2,175	JF	1,672	2,289	JF	1,435	BF	1,374	BF	2,556	806	806	JF	876	641	JF
2T4	3,897	2,053	JF	1,672	2,233	JF	1,446	BF	1,366	BF	2,625	674	674	JF	883	507	JF
1T44	3,959	2,055	JF	1,672	2,298	JF	1,464	BF	1,398	BF	2,701	673	673	JF	890	507	JF
3T4	4,762	2,680	JF	1,864	2,670	JF	1,898	BF	1,754	BF	3,368	1,263	1,263	BF	1,026	1,140	BF

(continued)

Table 6.3 (continued)

Specimen designation (total no of specimens: 51): Horizontal joint strength—V_{jRh} (kN); code specified transverse reinforcement—$A_{sv,c}$ (mm²); existing transverse reinforcement—$A_{sv,exp}$ (mm²); horizontal component of maximum sustained force by incline strut—$F_{jmax,h}$ (kN); horizontal component of compressive force corresponding to tensile force sustained by existing transverse reinforcement—$F_{j,h}$ (kN); failure mode—FM (joint failure—JF; beam failure—BF; experimentally established joint strength—$V_{j,exp}$ (kN)

Transverse reinforcement normally in the form of stirrups; suffix "d" indicates diagonal form of transverse reinforcement

Specimen	EC2–EC8			ACI 318			Vollume and Newman		Tsonos		Kotsovou and Mouzakis				Experimental results		
	V_{jRh}	$A_{sv,c}$	FM	V_{jRh}	$A_{sv,c}$	FM	V_{jRh}	FM	V_{jRh}	FM	$F_{jmax,h}$	$F_{j,h}$	V_{jRh}	FM	$V_{j,exp}$	$A_{sv,exp}$	FM
2T5	4,720	2,502	JF	1,864	2,537	JF	1,786	BF	1,704	BF	3,422	945	945	JF	1,018	792	BF
1T55	4,456	2,496	JF	1,864	2,316	JF	1,721	BF	1,564	BF	3,076	950	950	JF	1,023	792	BF
Karayannis et al. [35]																	
J0	851	543	JF	560	89	JF	337	BF	253	JF	418	–	–	JF	291	0	JF
JS	750	665	JF	496	86	BF	299	BF	249	JF	370	162	162	JF	288	101	JF
JX10	851	543	JF	560	89	BF	360	BF	253	JF	405	364	364	BF	283	157d	BF
JX12	851	543	JF	560	89	BF	400	BF	253	JF	401	525	401	BF	282	226d	BF
Karayannis et al. [36]																	
A1	536	181	JF	374	189	JF	108	BF	183	BF	257	169	169	BF	92	101	BF
A2	536	181	BF	374	251	JF	256	BF	196	BF	257	339	257	BF	92	201	BF
A3	536	181	BF	374	233	BF	279	BF	208	BF	257	508	257	BF	92	301	BF
B1	851	543	JF	560	226	JF	328	BF	239	JF	413	163	163	JF	281	101	JF
C2	851	521	JF	560	302	JF	386	BF	252	JF	380	483	380	BF	268	201	BF
C3	851	521	JF	560	339	JF	444	BF	265	JF-BF	380	724	380	BF	268	302	BF
C5	851	521	JF	560	377	BF	449	BF	292	BF	380	1,207	380	BF	269	503	BF
Chalioris et al. [37]																	
JA-S5	905	521	JF	581	405	BF	465	BF	297	BF	417	824	417	BF	267	503	BF
JA-X12	905	521	JF	581	405	JF	411	BF	220	JF	404	525	404	BF	266	226d	BF
JA-X14	905	521	JF	581	405	JF	458	BF	220	JF	400	714	400	BF	266	308d	BF
JB-S1	851	563	JF	560	563	JF	328	BF	239	JF	420	160	160	JF	280	101	JF
JB-X10	851	563	JF	560	563	JF	360	BF	213	JF	408	364	364	BF	274	157d	BF

(continued)

Table 6.3 (continued)

| Specimen designation (total no of specimens: 51) | Horizontal joint strength—V_{jRh} (kN); code specified transverse reinforcement—$A_{sv,c}$ (mm²); existing transverse reinforcement—$A_{sv,exp}$ (mm²); horizontal component of maximum sustained force by incline strut—$F_{jmax,h}$ (kN); horizontal component of compressive force corresponding to tensile force sustained by existing transverse reinforcement—$F_{j,h}$ (kN); failure mode—FM (joint failure—JF; beam failure—BF); experimentally established joint strength—$V_{j,exp}$ (kN). Transverse reinforcement normally in the form of stirrups; suffix "d" indicates diagonal form of transverse reinforcement |

Specimen	EC2–EC8			ACI 318			Vollume and Newman		Tsonos		Kotsovou and Mouzakis				Experimental results		
	V_{jRh}	$A_{sv,c}$	FM	V_{jRh}	$A_{sv,c}$	FM	V_{jRh}	FM	V_{jRh}	FM	$F_{jmax,h}$	$F_{j,h}$	V_{jRh}	FM	$V_{j,exp}$	$A_{sv,exp}$	FM
JB-X12	851	563	JF	560	563	JF	401	BF	213	JF	403	525	403	BF	273	226d	BF
JCa-X10	175	174	JF	151	145	JF	121	BF	32	JF	73	365	73	JF-BF	73	157	JF-BF
JCa-S1	175	174	JF	151	145	JF	121	BF	53	JF	79	164	73	JF-BF	73	101	JF-BF
JCa-S1-X10	175	174	BF	151	145	BF	121	BF	53	JF	73	533	73	JF-BF	73	101s, 157d	JF-BF
JCa-S2	175	174	BF	151	195	BF	121	BF	63	JF	79	328	73	JF-BF	73	201	JF-BF
JCa-S2-X10	175	174	BF	151	195	BF	121	BF	63	JF	73	701	79	BF	73	201s, 157d	BF
JCb-X10	193	260	JF	160	162	JF	128	BF-JF	36	JF	79	365	79	JF-BF	79	157d	JF-BF
JCb-S1	193	260	JF	160	162	JF	128	JF	57	JF	77	165	77	JF	79	101	JF
JCb-S1-X10	193	260	JF-BF	160	162	BF	128	BF	57	JF	79	535	79	JF-BF	79	101s, 157d	JF-BF
JCb-S2	193	260	JF	160	217	JF	128	BF-JF	67	JF	77	331	77	JF	79	201	JF
JCb-S2-X10	193	260	BF	160	217	BF	128	BF	67	JF	79	705	79	JF-BF	79	201s, 157d	JF-BF
Tsonos [38]																	
A1	334	268	BF	393	542	JF	293	BF	209	BF	237	769	237	BF	149	483	BF
E1	132	320	JF	311	341	BF	233	BF	154	JF	156	921	156	JF	223	483	JF
E2	334	260	BF	393	521	JF	293	BF	211	BF	286	782	237	BF	146	483	BF
G1	132	405	JF	311	246	BF	233	BF	145	JF	156	615	156	JF	230	343	JF
Kotsovou and Mouzakis [18]																	

(continued)

Table 6.3 (continued)

Specimen designation (total no of specimens: 51)	Horizontal joint strength—V_{jRh} (kN); code specified transverse reinforcement—$A_{sv,c}$ (mm²); existing transverse reinforcement—$A_{sv,exp}$ (mm²); horizontal component of maximum sustained force by incline strut—$F_{jmax,h}$ (kN); horizontal component of compressive force corresponding to tensile force sustained by existing transverse reinforcement—$F_{j,h}$ (kN); failure mode—FM (joint failure—JF; beam failure—BF; experimentally established joint strength—$V_{j,exp}$ (kN)
	Transverse reinforcement normally in the form of stirrups; suffix "d" indicates diagonal form of transverse reinforcement

	EC2–EC8			ACI 318			Vollume and Newman		Tsonos		Kotsovou and Mouzakis				Experimental results		
	V_{jRh}	$A_{sv,c}$	FM	V_{jRh}	$A_{sv,c}$	FM	V_{jRh}	FM	V_{jRh}	FM	$F_{jmax,h}$	$F_{j,h}$	V_{jRh}	FM	$V_{j,exp}$	$A_{sv,exp}$	FM
S1	1,300	1,478	BF	884	390	BF	660	BF	767	BF	539	2,358	539	JF	632	1,544	JF
S2	1,300	1,478	BF	884	390	BF	660	BF	767	BF	539	2,358	539	JF	624	1,544	JF
S3	1,300	1,478	BF	884	390	BF	660	BF	685	BF	539	1,769	539	JF	624	1,158	JF
S4	1,300	1,478	BF	884	390	BF	660	JF	526	JF	539	1,319	539	JF	626	452(s) 314(d)	JF
S5	1,300	1,478	BF	884	390	BF	660	JF	526	JF	538	1,659	538	JF	617	452(s) 340(d)	JF
Kotsovou and Mouzakis [25]																	
S6	2,459	1,478	BF	1,375	465	BF	1,245	BF	846	BF	1,291	1,732	1,291	BF	722	402(s) 616(d)	BF
S9	2,459	1,478	BF	1,375	425	BF	1,259	BF	960	BF	1,291	1,174	1,174	BF	722	943	BF
S10	1,300	739	BF	884	476	BF	660	BF	496	BF	532	1,521	532	BF	355	343(s) 452(d)	BF
S11	2,459	1,478	BF	1,375	425	BF	1,259	BF	960	BF	1,291	1,174	1,174	BF	722	943	BF

Table 6.4 Comparison of predictions with published experimental information on beam-column sub-assemblages characterised by failure of the joint

Horizontal Joint Strength—V_{jRh} (kN); code specified transverse reinforcement—A_{svc} (kN); existing transverse reinforcement—A_{svc} (mm²); existing transverse reinforcement—$A_{sv,exp}$ (mm²); horizontal component of maximum sustained force by incline strut—$F_{jmax,h}$ (kN); horizontal component of compressive force corresponding to tensile force sustained by existing transverse reinforcement—$F_{j,h}$ (kN); joint failure—JF; beam failure—BF; experimentally established joint strength—$V_{j,exp}$ (kN)

Transverse reinforcement normally in the form of stirrups; suffix "d" indicates diagonal form of transverse reinforcement

Specimen Designation (total no of specimens: 26)	EC2–EC8			ACI 318			Vollum &Newman		Tsonos		Kotsovou & Mouzakis				Experimental results		
	V_{jRh}	A_{svc}	FM	V_{jRh}	A_{svc}	FM	V_{jRh}	FM	V_{jRh}	FM	$F_{jmax,h}$	$F_{j,h}$	V_{jRh}	FM	$V_{j,exp}$	$A_{sw,exp}$	FM
Kurose [32]																	
S41	949	1,791	JF	688	676	JF	435	BF	299	JF	397	343	343	JF	408	382	JF
S42	949	1,791	JF	688	676	BF	547	BF	333	JF	397	685	397	JF	402	763	JF
U41L	1,029	1,791	JF	720	742	JF	450	BF	325	JF	439	343	343	JF	413	382	JF
Eshani and Wight [20]																	
1B	945	956	JF	806	1,137	JF	617	JF	438	JF	475	1,182	475	JF	700	865	JF
2B	946	935	JF	822	1,083	JF	668	JF	444	JF	538	1,149	538	JF	707	865	JF
3B	1,102	953	BF	889	1,557	JF	681	BF	548	BF	600	1,763	600	JF	671	1,298	JF
5B	929	1,762	JF	892	590	BF	736	BF	486	JF	561	1,164	561	JF	831	865	JF
Eshani and Alameddine [33]																	
HL8	1,962	2,732	JF	1,288	2,248	JF	1,207	BF	926	BF	1,166	2,029	1,166	JF	1,185	1,520	JF
HH8	1,962	2,732	JF	1,288	2,248	JF	1,207	BF	1,011	BF	1,166	3,045	1,166	JF	1,172	2,280	JF
Tsonos et al. [21]																	
S6	464	703	JF	381	451	JF	285	JF	182	JF	225	437	225	JF	314	302	JF
S6'	407	692	JF	357	369	JF	267	JF	165	BF	165	497	165	JF	230	302	JF
X6	381	689	JF	345	396	BF	258	BF	157	JF	166	1,090	166	JF	298	302s 308d	JF
Chutarat and Aboutaha [34]																	
1	2,074	2,316	JF	1,350	668	BF	1,154	BF	976	JF	769	1,711	769	JF	986	2,027	JF
Hwang et al. [22]																	
OTO	4,091	2,058	JF	1,672	2,138	JF	1,408	JF	1,202	BF	2,532	–	–	JF	896	0	JF

(continued)

Table 6.4 (continued)

Horizontal Joint Strength—V_{jRh} (kN); code specified transverse reinforcement—$A_{sv,c}$ (mm²); existing transverse reinforcement—$A_{sv,exp}$ (mm²); horizontal component of maximum sustained force by incline strut—$F_{jmax,h}$ (kN); horizontal component of compressive force corresponding to tensile force sustained by existing transverse reinforcement—$F_{j,h}$ (kN); failure mode—FM (joint failure—JF; beam failure—BF); experimentally established joint strength—$V_{j,exp}$ (kN)

Transverse reinforcement normally in the form of stirrups; suffix "d" indicates diagonal form of transverse reinforcement

Specimen Designation (total no of specimens: 26)	EC2-EC8			ACI 318			Vollume & Newman		Tsonos		Kotsovou & Mouzakis				Experimental results		
	V_{jRh}	$A_{sv,c}$	FM	V_{jRh}	$A_{sv,c}$	FM	V_{jRh}	FM	V_{jRh}	FM	$F_{jmax,h}$	$F_{j,h}$	V_{jRh}	FM	$V_{j,exp}$	$A_{sv,exp}$	FM
3T3	3,910	2,175	JF	1,672	2,289	JF	1,435	BF	1,374	BF	2,556	806	806	JF	876	641	JF
2T4	3,897	2,053	JF	1,672	2,233	JF	1,446	BF	1,366	BF	2,625	674	674	JF	883	507	JF
1T44	3,959	2,055	JF	1,672	2,298	JF	1,464	BF	1,398	BF	2,701	673	673	JF	890	507	JF
Karagiannis et al. [35]																	
J0	851	543	JF	560	89	JF	337	BF	253	JF	418	–	–	JF	291	0	JF
JS	750	665	JF	496	86	BF	299	BF	249	JF	370	162	162	JF	288	101	JF
Karagiannis et al. [36]																	
B1	851	543	JF	560	226	JF	328	BF	239	JF	413	163	163	JF	281	101	JF
Chalioris et al. [37]																	
JB-S1	851	563	JF	560	563	JF	328	BF	239	JF	420	160	160	JF	280	101	JF
JCa-X10	175	174	JF	151	145	JF	121	BF	32	JF	73	365	73	JF-BF	73	157	JF-BF
JCa-S1	175	174	JF	151	145	JF	121	BF	53	JF	79	164	73	JF-BF	73	101	JF-BF
JCa-S1-X10	175	174	BF	151	145	BF	121	BF	53	JF	73	533	73	JF-BF	73	101s 157d	JF-BF
JCa-S2	175	174	BF	151	195	BF	121	BF	63	JF	79	328	73	JF-BF	73	201	JF-BF
JCb-X10	193	260	JF	160	162	BF-JF	128	BF	36	JF	79	365	79	JF-BF	79	157d	JF-BF
JCb-S1	193	260	JF	160	162	JF	128	BF	57	JF	77	165	77	JF	79	101	JF
JCb-S1-X10	193	260	BF	160	162	BF	128	BF	57	JF	79	535	79	JF-BF	79	101s 157d	JF-BF
JCb-S2	193	260	JF	160	217	BF-JF	128	BF	67	JF	77	331	77	JF	79	201	JF

(continued)

Table 6.4 (continued)

Horizontal Joint Strength—V_{jRh} (kN); code specified transverse reinforcement—$A_{sv,c}$ (mm²); existing transverse reinforcement—$A_{sv,exp}$ (mm²); horizontal component of maximum sustained force by incline strut—$F_{jmax,h}$ (kN); horizontal component of compressive force corresponding to tensile force sustained by existing transverse reinforcement—$F_{j,h}$ (kN); failure mode—FM (joint failure—JF; beam failure—BF); experimentally established joint strength—$V_{j,exp}$ (kN)

Transverse reinforcement normally in the form of stirrups; suffix "d" indicates diagonal form of transverse reinforcement

Specimen Designation (total no of specimens: 26)	EC2–EC8			ACI 318			Vollume &Newman		Tsonos		Kotsovou & Mouzakis				Experimental results		
	V_{jRh}	$A_{sv,c}$	FM	V_{jRh}	$A_{sv,c}$	FM	V_{jRh}	FM	V_{jRh}	FM	$F_{jmax,h}$	$F_{j,h}$	V_{jRh}	FM	$V_{j,exp}$	$A_{sv,exp}$	FM
JCb-S2-X10193	132	260	BF	160	217	BF	128	BF	67	JF	79	705	79	JF-BF	79	201s 157d	JF-BF
Tsonos [38]]																	
E1	132	320	JF	311	341	BF	233	BF	154	BF	156	921	156	JF	223	483	JF
G1	132	405	JF	311	246	BF	233	BF	145	BF	156	615	156	JF	230	343	JF
Kotsovou and Mouzakis [18]																	
S1	1,300	1,478	BF	884	390	BF	660	BF	767	BF	539	2,358	539	JF	632	1,544	JF
S2	1,300	1,478	BF	884	390	BF	660	BF	767	BF	539	2,358	539	JF	624	1,544	JF
S3	1,300	1,478	BF	884	390	BF	660	BF	685	BF	539	1,769	539	JF	624	1,158	JF
S4	1,300	1,478	BF	884	390	BF	660	JF	526	JF	539	1,319	539	JF	626	452(s) 314(d)	JF
S5	1,300	1,478	BF	884	390	BF	660	JF	526	JF	538	1,659	538	JF	617	452(s) 340(d)	JF

successfully predict the mode of failure in 27 and 18, respectively, cases, whereas those proposed by Vollum and Newman [23] and Tsonos [14] successfully predict the mode of failure in 7 and 30, respectively, cases; in the remainder, joint capacity is overestimated, since failure due to the formation of a plastic hinge at the beam-joint interface is predicted to occur before the joint capacity is exhausted. In fact, EC2 [7]–EC8 [17] is found to overestimate joint capacity in cases for which the existing amount of transverse reinforcement is either less (specimens S3, S4 and S5) or more (specimens 3B, S1 and S2) than the code specified value, with ACI318 [6] producing similar predictions, but, always in cases where the amount of the existing transverse reinforcement is more than the code specified value. On the other hand, the formulae proposed in the preceding section successfully predict the mode of failure in all 37 cases included in Table 6.4.

Table 6.5 provides an indication of the closeness with which the expressions investigated predict joint capacity. The latter is expressed in a non-dimensional form by dividing the predicted values with their experimental counterparts. As regards the EC2 [7]–EC8 [17] and the ACI [6] predictions of joint failure due to excessive cracking, it is interesting to note in the table that, with the exception of specimens E1 and G1 for the case of EC2 [7]–EC8 [17], the predicted behaviour is based on the codes' requirement for a specific amount of transverse reinforcement which is not fulfilled. Moreover, as the code adopted formulae for joint capacity do not allow for the effect of transverse reinforcement, the assessment of joint capacity through the use of the existing reinforcement is not possible in this case. On the other hand, for the small number of cases for which the formulae proposed by Vollum and Newman [23] successfully predict joint failure due to excessive cracking, the mean value of the predicted joint capacity is only slightly smaller than its experimental counterpart, whereas that predicted by the formulae proposed by Tsonos [14] underestimates the experimental value by about 25 %; in the latter case, however, the mean value is obtained from a significantly larger number of successful predictions. Finally, in contrast with the formulae already discussed, those proposed in the preceding section are found to produce realistic predictions in all cases investigated, with their mean value underestimating joint capacity by an amount of the order of 20 %.

It appears from the above, therefore, that the Kotsovou and Mouzakis [25] and the Tsonos formulae [14] produce by far the better predictions of joint behaviour as regards to both the mode of failure and the deviation of the predicted from the experimentally-established joint capacity. Of the above formulae, the former are found the most effective for the cases investigated and this is considered to primarily reflect the validity of the assumption underlying the derivation of the formulae that the diagonal strut is the sole mechanism of force transfer.

6.5 Concluding Remarks

Any structural configuration comprising beam-column elements can be visualised as an assemblage of simply supported beams (extending between consecutive points of zero bending moments) or cantilevers (extending between a

Table 6.5 Comparison of predicted joint strength with published experimental information

Specimen designation (total no of specimens: 26)	Joint Strength—V_{jRh} (kN); experimentally established joint strength—$V_{j,exp}$ (kN); lack of specified transverse reinforcement—lstr; without transverse reinforcement—wtr				
	EC2–EC8	ACI 318	Vollume and Newman	Tsonos	Kotsovou and Mouzakis
	$V_{jRh}/V_{j,exp}$				
Kurose [33]					
S41	lstr	lstr	–	0.73	0.84
S42	lstr	–	–	0.83	0.99
U41L	lstr	lstr	–	0.78	0.83
Eshani and Wight [20]					
1B	lstr	lstr	0.88	0.63	0.68
2B	lstr	lstr	0.94	0.63	0.76
3B	–	lstr	–	0.82	0,89
5B	lstr	–	0.89	0.58	0.68
Eshani and Alameddine [33]					
HL8	lstr	lstr	–	0.78	0.98
HH8	lstr	–	–	0.86	0.99
Tsonos et al. [21]					
S6	lstr	lstr	0.91	0.58	0.72
S6'	lstr	lstr	–	0.72	0.72
X6	lstr	–	0.87	0.53	0.56
Chutarat and Aboutaha [34]					
I	lstr	–	–	0.99	0.78
Hwang et al. [22]					
0T0	lstr	wtr	–	–	–
3T3	lstr	lstr	–	–	0.92
2T4	lstr	lstr	–	–	0.76
1T44	lstr	lstr	–	–	0.76
Karagiannis et al. [35]					
J0	wtr	wtr	–	0.87	–
JS	lstr	–	–	0.86	0.56
Karagiannis et al. [36]					
B1	lstr	lstr	–	0.85	0.58
Chalioris et al. [37]					
JB-S1	lstr	lstr	–	0.85	0.57
JCa-X10	lstr	lstr	–	0.44	1
JCa-S1	lstr	lstr	–	0.73	1
JCa-S1-X10	lstr	lstr	–	0.73	1
JCa-S2	–	–	–	0.86	1
JCb-X10	lstr	lstr	–	0.46	1
JCb-S1	lstr	lstr	–	0.72	0.97
JCb-S1-X10	–	–	–	0.72	1
JCb-S2	lstr	lstr	–	0.85	1
JCb-S2-X10	–	–	–	0.85	1
Tsonos [38]					
E1	0.59	–	–	0.69	0.7

Specimen designation (total no of specimens: 26)	Joint Strength—V_{jRh} (kN); experimentally established joint strength—$V_{j,exp}$ (kN); lack of specified transverse reinforcement—lstr; without transverse reinforcement—wtr				
	EC2–EC8	ACI 318	Vollume and Newman	Tsonos	Kotsovou and Mouzakis
	$V_{jRh}/V_{j,exp}$				
G1	0.57	–	–	0.63	0.69
Kotsovou and Mouzakis [18]					
S1	–	–	–	–	0.85
S2	–	–	–	–	0.86
S3	–	–	–	–	0.86
S4	–	–	1.05	0.84	0.86
S5	–	–	1.07	0.85	0.87
Predictions of JF (mean value/SD)	0.58/0.01		0.94/0.08	0.74/0.13	0.83/0.15

beam-column joint and its adjacent point of zero bending moment). The stress conditions in the region of a beam-to-beam or a column-to-column connection resemble those developing at the adjacent ends of beam or column elements where the one suspends from the other through the development of a transverse tie, whereas a beam-to-column connection functions as a diagonal strut developing within the common portion of the intersecting beam and column elements.

In the absence of axial force, a beam-to-beam or a column-to-column connection fails when the strength of the transverse tie is exhausted; on the other hand, in the presence of axial compression, failure is linked with the change in the path of the compressive force through the joint in a manner similar to that for the cases of beam behaviour of types II and III, as discussed in Chaps. 3 and 4.

For the case of a beam-to-column connection, the strength of the diagonal strut may be assessed as for the case of beam behaviour of type IV; however, in this case, the provision of transverse reinforcement is essential in order to prevent excessive cracking and progressive loss of load-carrying capacity in the case of load reversals induced by seismic loading.

References

1. Kotsovos GM, Kotsovos DM, Kotsovos MD, Kounadis A (2011) Seismic design of structural concrete walls: an attempt to reduce reinforcement congestion. Mag Concr Res 63(4):235–245
2. Kotsovos MD, Pavlovic MN (1999) Ultimate limit-state design of concrete structures: a new approach. Thomas Telford, London, p 164
3. Kotsovos MD (1979) Fracture of concrete under generalised stress. Mater Struct, RILEM 12(72):151–158
4. Kotsovos MD, Newman JB (1981) Fracture mechanics and concrete behaviour. Mag Concr Res 33(115):103–112

5. Kotsovos GM, Kotsovos MD (forthcoming) Effect of axial compression on shear capacity of linear RC members without transverse reinforcement. Mag Concr Res
6. American Concrete Institute (2002) Building code requirements for structural concrete (ACI 318-02) and commentary (ACI 318R-02)
7. EN 1992-1 (2004) Eurocode 2: design of concrete structures—part 1-1: general rules and rules for buildings
8. De Cossio RD, Siess CP (1960) Behaviour and strength in shear of beams and frames without web reinforcement. ACI J 56:695–735
9. Gupta PR, Collins MP (2001) Evaluation of shear design procedures for reinforced concrete members under axial compression. ACI Struct J 98(4):537–547
10. Paulay T, Park R, Priestey MJN (1978) Reinforced concrete beam—column joints under seismic actions. ACI J 75(11):585–593
11. Paulay T (1989) Equilibrium criteria for reinforced concrete beam—column joints. ACI Struct J 86(6):635–643
12. Leon RT (1990) Shear strength and hysteric behaviour of interior beam—column joints. ACI Struct J 87(1):3–11
13. Paulay T, Priestley MJN (1992) Seismic design of reinforced concrete and masonry buildings. Willey, New York
14. Tsonos AG (2006) Cyclic load behaviour of reinforced concrete beam—column subassemblages designed according to modern codes. Eur Earthq Eng 3:3–21
15. Paulay T, Paulay T (1975) Reinforced concrete structures. Wiley, New York 769
16. NZS 3101 (1995) Concrete structures standard. Standards Association of New Zealand, Wellington
17. EN 1998-1 (2004) Eurocode 8: design of structures for earthquake resistance—part 1: general rules, seismic actions and rules for buildings
18. Kotsovou GM, Mouzakis H (2011) Seismic behaviour of RC external joints. Mag Concr Res 33(4):247–264
19. Ehsani MR, Wight JK (1985) Effect of transverse beams and slab on behaviour of reinforced concrete beam—to—column connections. ACI J 82(2):188–195
20. Ehsani MR, Wight JK (1985) Exterior reinforced concrete beam—to—column connections subjected to earthquake—type loading. ACI J 82(4):492–499
21. Tsonos AG, Tegos IA, Penelis G Gr (1992) Seismic resistance of type 2 exterior beam-column joints with inclined bars. ACI Struct J 89(1):3–12
22. Hwang S, Lee H-J, Liao T-F, Wang KC, Tsai H-H (2005) Role of hoops on shear strength of reinforced concrete beam—column joints. ACI Struct J 102(3):445–453
23. Vollum RL, Newman JB (1999) Strut and tie models for the analysis/design of external beam-column joints. Mag Concr Res 51(6):415–425
24. Hegger J, Sherif A, Roeser W (2003) Nonseismic design of beam-column joints. ACI Struct J 100(5):654–664
25. Kotsovou G, Mouzakis H (2012) Seismic design of RC external beam-column joints. Bull Earthq Eng 10(2):645–677
26. Zhang L, Jirsa JO (1982) A study of the shear behaviour of reinforced concrete beam-column joints. RMFSEL report no. 82-1, Phil. M. Ferguson Structural Engineering Laboratory, University of Texas at Austin, Feb 1982
27. Sarsam KF, Phipps ME (1985) The shear design of in situ reinforced concrete beam-column joints subjected to monotonic loading. Mag Concr Res 37(130):16–28
28. Design and detailing of concrete structures for fire resistance (1978) Design and detailing of concrete structures for fire resistance. Interim guidance by a joint committee of the institution of structural engineers and the concrete society. The Institution of Structural Engineers, April 1978, p 59
29. Kotsovos MD (1988) Design of reinforced concrete deep beams. Struct Eng 66(2):28–32
30. Kotsovou GM, Kotsovos MD (2013) Towards improving the design of RC exterior beam-column joints. Struct Eng 91(3):40–50

31. Hanson NW, Connor HW (1967) Seismic resistance of reinforced concrete beam-column joints. J Struct Div, ASCE 93(ST5):533–559
32. Kurose Y (1987) Recent studies on reinforced concrete beam column joints in Japan. PMFSEL report no. 87-8, Phil M. Ferguson Structural Engineering Laboratory, Department of Civil Engineering University of Texas at Austin, Dec 1987
33. Ehsani MR, Alameddine F (1991) High-strength RC connections subjected to inelastic cyclic loading. J Struct Eng, ASCE 117(3):829–850
34. Chutarat N, Aboutaha RS (2003) Cyclic responce of exterior reinforced concrete beam-column joints reinforced with headed bars-experimental investigation. ACI Struct J 100(2):259–264
35. Karayannis CG, Chalioris CE, Sirkelis GM (2006) Exterior RC beam-column joints with diagonal reinforcement. In: Proceedings of the 15th concrete congress, Hellenic Technical Chamber, Alexandroupolis, Greece, vol. A, pp 368–377 (in Greek)
36. Karayannis CG, Chalioris CE, Sirkelis GM (2008) Local retrofit of exterior RC beam-column joints using thin RC jackets—an experimental study. Earthq Eng Struct Dynam 37:727–746
37. Chalioris CE, Favvata MJ, Karayannis CG (2008) Reinforced concrete beam-column joints with crossed inclined bars under cyclic deformations. Earthq Eng Struct Dynam 37:881–897
38. Tsonos AG (2007) Cyclic load behavior of Reinforced Concrete beam-column subassemblages of modern structures. ACI Struct J 104(4):468–478

Chapter 7
Earthquake-Resistant Design

7.1 Introduction

From both practical experience and published experimental evidence, it becomes clear that the methods adopted by current codes for the design of earthquake-resistant RC structures have two significant shortcomings: Not only do they lead to reinforcement congestion which may cause difficulties in concreting and often incomplete compaction [1], but, also, in spite of the large amount of reinforcement specified, they have been found unable to always prevent the brittle types of failure which they are widely considered to safeguard against [2].

This chapter presents experimental evidence which shows that the application of the methods proposed in Chaps. 3, 4, 5 and 6 for designing earthquake-resistant RC structures can minimize, if not eliminate, the above shortcomings. This evidence has been obtained from published work on the behaviour of structural members such as beam-column elements [3], structural walls [1], and beam-column joint sub-assemblages [4].

7.2 Beam-Column Elements

The structural performance requirements of current codes, which an RC beam or column must comply with, include safeguarding a specified load-carrying capacity and adequate ductility (i.e. ability of sustaining post-peak displacements at least as large as a specified multiple of the displacement at yield). In order to safeguard solutions that satisfy the latter of the above requirements, current codes for the design of earthquake-resistant RC structures specify additional amounts of stirrup reinforcement in regions (under the action of the largest bending moment combined with the largest shear force) classified as 'critical' [5]. And yet, not only does the provision of such reinforcement lead to reinforcement congestion [1], but also its inclusion has been found insufficient to safeguard the code specified structural performance [2].

M. D. Kotsovos, *Compressive Force-Path Method*, Engineering Materials,
DOI: 10.1007/978-3-319-00488-4_7, © Springer International Publishing Switzerland 2014

The application of the proposed method for the earthquake-resistant design of beam-column elements has formed the subject of a number of publications [3, 6–8]. The work reported is experimental and describes the results obtained from tests on specimens under load mimicking seismic action. The results discussed in what follows are extracted from Ref. [3] and are typical of those presented in the above publications where full details of the work carried out can be found.

7.2.1 Experimental Details

The specimens investigated are simply-supported beam-column elements such as that shown in Fig. 7.1, the latter also showing the specimens' cross-sectional characteristics, the load arrangement and the relevant bending-moment and shear-force diagrams. It is interesting to note in the figure that portions AB and BC of the structural element are subjected to internal actions similar to those of the portion of a column between its point of contraflexure (inflection) and one of its ends. From the geometric characteristics of the above portions indicated in the figure, the values of a_v/d for portions AB and BC are approximately equal to 3.5.

From Fig. 7.1, it can be seen that the specimens have a span of 1,950 mm and a 300 mm high $\times$ 150 mm wide cross section. The longitudinal reinforcement comprises deformed 14 mm diameter bars with average values of yield stress and

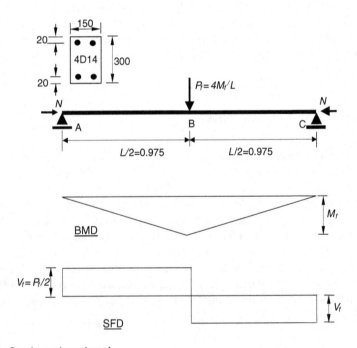

Fig. 7.1 Specimens investigated

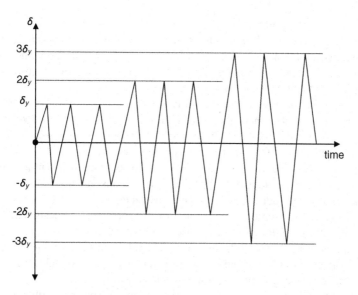

Fig. 7.2 Loading history adopted for the cyclic tests

strength $f_y = 540$ MPa and $f_u = 640$ MPa, respectively. The stirrups are made from 6 mm diameter mild steel bars with a yield stress $f_{yv} = 300$ MPa. The uniaxial cylinder compressive strength of concrete at the time of testing of the specimens is $f_c = 60$ MPa at an age of around one year.

The specimens are subjected to sequential loading comprising axial (N) and transverse (P) components, as indicated in Fig. 7.1. The axial load is applied first; it increases to a predefined value $N \approx aN_u = af_cbh$ (where N_u is the maximum value of N that can be sustained by the specimen in concentric compression, b and h the cross-sectional dimensions (width and height, respectively), whereas the values of a selected for the tests are 0, 0.15, 0.2 and 0.25), where it is maintained constant during the subsequent application of P. The latter force is applied at mid span in a cyclic manner inducing progressively increasing displacements in opposite directions as shown in Fig. 7.2. Full details of the experimental set up used for the tests are provided elsewhere [3].

7.2.2 Specimen Design

For purposes of comparison two of the specimens are designed in compliance with EC2 [9] and EC8 [5] and the remainder in accordance with the proposed method, with all safety factors being taken equal to 1. In all cases, it is assumed that load-carrying capacity is reached when the mid cross-section of the specimens attains its flexural capacity, the latter condition being referred to henceforth as *plastic-hinge* formation. Flexural capacity, denoted as M_f, is calculated as described in Sect. 3.3.1 for a value of $N \approx abhf_c$, with a, as discussed earlier, obtaining values equal to 0, 0.15, 0.2 and 0.25 corresponding to values of N equal to 0, 400, 500 and 675 kN, respectively. Henceforth, the above structural elements are referred to by using a two-part denomination arranged in sequence to denote whether they are designed in accordance with the proposed (CFP) or the code (EC) method and the value of N (0, 400, 500, and 675) applied.

Using M_f, the specimen's load-carrying capacity $P_f = 4M_f/L$ (where L the specimen's span) and, hence, the corresponding shear force $V_f = P_f/2$ is easily calculated. The values of M_f and P_f for each of the specimens tested are given in Table 7.1 together with the experimentally-established values of the load-carrying capacity P_{exp}. The table also includes the values of bending moment M_y and load P_y which correspond at the yielding of the cross sections; the latter values are used as discussed later in order to assess the ductility factor μ of the specimens tested.

The transverse reinforcement (stirrups) of the specimens is designed either in accordance with the CFP method or in compliance with the earthquake-resistant design clauses of EC2 and EC8. The calculated values of the spacing of the 6 mm diameter bars used to form the stirrup reinforcement are given in Table 7.2, whereas the location of the stirrups indicated in the table is shown in Fig. 7.3.

From Table 7.2, it is interesting to note the densely spaced stirrups within the "critical region" (which, as shown in Fig. 7.3, extends to a distance of 300 mm

Table 7.1 Calculated values of bending moment M_y and corresponding force P_y at yield, flexural capacity M_f and corresponding load-carrying capacity P_f, and experimentally-established values of load-carrying capacity P_{exp}

Specimens	Design method	N (kN)	M_y (kNm)	P_y (kN)	M_f (kNm)	P_f (kN)	P_{exp} (kN)
CFP-0	CFP	0	40.5	83.1	46.7	95.7	102.5
EC-0	EC	0	40.5	83.1	46.7	95.7	100.0
CFP-400	CFP	400	85.0	174.3	91.8	188.3	203.4
EC-400	EC	400	85.0	174.3	91.8	188.3	201.4
CFP-500	CFP	500	88.5	181.5	101.0	207.1	224.4
CFP-675	CFP	675	90.0	184.6	118.0	242.1	264.7

Table 7.2 Spacing of 6 mm diameter stirrups

Spacing (mm)		
Specimen	Critical region (see Fig. 7.3)	Region 1[a]
CFP-0	65	140
EC-0	75	150
CFP-400	75	95
EC-400	50	110
CFP-500	80	85
EC-500[b]	30	100
CFP-675	85	75
EC-675[b]	15	90

[a]For the CFP specimens, the stirrups required in region 1 are only placed within a length equal to $2d$ extending symmetrically about location 1 in Fig. 3.7 (location of change in CFP)
[b]Specimens not manufactured due unrealistically small spacing of stirrups within critical region

Fig. 7.3 Regions of transverse reinforcement indicated in Table 7.2

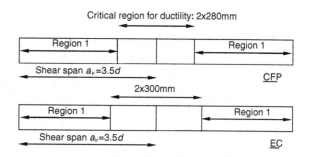

on either side of the mid span cross section) specified by the Codes in order to provide confinement to concrete. Such spacing [resulting from expressions 5.15 in EC8 (clause 5.4.3.2.2)] is considered to safeguard ductile specimen behaviour. It should be noted, however, that, for the specimens investigated, expressions 5.15 results in an applicable stirrup spacing of the order of 50 mm only for the specimens subjected to an axial force $N = 400$ kN; in all other cases, they resulted in unrealistically small values (significantly smaller than 50 mm). As a result, only two of the specimens tested are designed in compliance with the code provisions: one subjected to the combined action of cyclic transverse loading and a constant $N = 400$ kN and the other subjected only to cyclic transverse loading, with the latter one having stirrup spacing equal to $b/2 = 75$ mm, which is the maximum code specified value [see expression 5.18 in EC8 (clause 5.4.3.2.2(11)].

As discussed in Sect. 4.3.1, in contrast with the code reasoning behind the calculation of the stirrups within the critical regions, the CFP method specifies a significantly smaller amount of such reinforcement not for providing confinement to concrete, but in order to sustain the transverse tensile stresses developing within the compressive zone as a result of stress redistribution due to the loss of bond between concrete and the flexural reinforcement in the regions of the shear span subjected to the largest bending moment.

On the other hand, within the remainder of the shear spans, transverse reinforcement designed in compliance with the code requirements (see clauses 6.2 and 9.5.3 in EC2) is considered to prevent shear failure of the specimens before their flexural capacity is exhausted. This reasoning is also in conflict with that underlying the design of the transverse reinforcement in accordance with the proposed method: in the latter case, as discussed in Sect. 4.3.1, transverse reinforcement is designed to sustain the tensile force developing in the region of change in the direction of the path of the compressive force developing on account of bending.

7.2.3 Results of Tests

The main results of the work are given in Figs. 7.4, 7.5, 7.6, 7.7 and Tables 7.1 and 7.3. Figures 7.4, 7.5 7.6 show the results obtained in the form of load–deflection curves. The load–deflection curves in Figs. 7.4 and 7.5 describe the specimens'

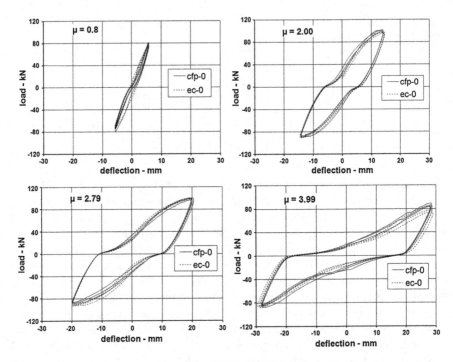

Fig. 7.4 Hysteretic behaviour of specimens CFP-0 and EC-0 under cyclic loading corresponding to ductility ratios $\mu = 0.8$, 2.00, 2.79, and 3.99

hysteretic behaviour under loading cycles to displacement values corresponding to particular ductility ratios for the cases of $N = 0$ ($a = 0$) and $N = 400$ kN ($a = 0.15$), respectively; moreover, for comparison purposes, the figures include the results obtained both for the specimens designed in compliance with the code requirements and for those designed in accordance with the proposed design method. On the other hand, Fig. 7.6a–d show the full load–deflection curves of the specimens designed in accordance with the CFP method and tested under the combined action of cyclic transverse loading and a constant axial force equal to $N = 0$ kN ($a = 0$), $N = 400$ kN ($a = 0.15$), $N = 500$ kN ($a = 0.2$) and $N = 675$ kN ($a = 0.25$). Typical modes of failure of the specimens are shown in Fig. 7.7 and, finally, the experimentally established values of load-carrying capacity are given in Table 7.1, whereas Table 7.3 includes the assessed (as described below) values of the deflection at nominal yield (δ_{yn}), the measured values of the maximum sustained deflection (δ_{sust}) and the values of the ductility ratio at the hysteretic loops corresponding to the maximum sustained displacement (μ_{sust}).

Figure 7.6a–d also include the location of the nominal yield point used for assessing the specimen ductility ratio μ_{sust}. The location of this point is determined as follows:

(a) The section bending moment at yield, M_y (assessed by assuming that yielding occurs when either the concrete strain at the extreme compressive fibre attains

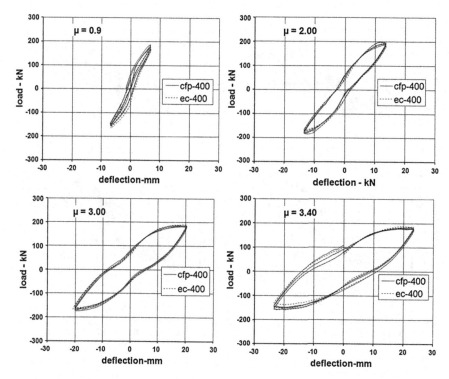

Fig. 7.5 Hysteretic behaviour of specimens CFP-400 and EC-400 under cyclic loading corresponding to ductility ratios $\mu = 0.9$, 2.0, 3.0, and 3.4

a value of 0.002 or the tension reinforcement yields), and the section flexural capacity, M_f, are first calculated.

(b) Using the values of M_y and M_f derived in (a), the corresponding values of the transverse load at yield, $P_y = M_y/a_v$, and at flexural capacity, $P_f = M_f/a_v$, are obtained from the equilibrium equations, with a_v (= 0.975 m) being the distance of the point of application of the applied load from the nearest support (see Fig. 7.1).

(c) In Fig. 7.6a–d, lines are drawn through the points of the load–displacement curves at $P = 0$ and $P = P_y$. These lines are extended to the load level P_f, which is considered to define the nominal yield point, and corresponds to displacement δ_{yn}, the later being used to calculate the ductility ratios $\mu_{sust} = \delta_{sust}/\delta_{yn}$ in Table 7.3.

7.2.4 Discussion of the Results

Figures 7.4 and 7.5 indicate that, for the $N = 0$ and $N = 400$ kN, the hysteretic response of both types of specimens investigated is almost identical. It appears, therefore, that the significantly denser stirrup spacing specified by the codes within

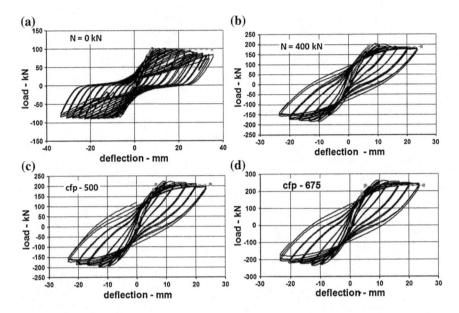

Fig. 7.6 Load-deflection curves of specimens designed in accordance with the proposed method under cyclic loading combined with an axial force equal to (**a**) 0, (**b**) 400 kN, (**c**) 500 kN, and (**d**) 675 kN

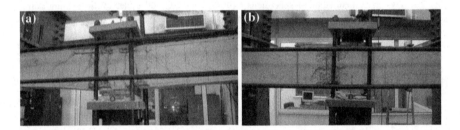

Fig. 7.7 Typical failure modes of the specimens tested under cyclic loading combined with (**a**) N = 0 and (**b**) N ≠ 0

Table 7.3 Values of displacements at nominal yield (δ_{yn}), maximum sustained displacements (δ_{sust}) and corresponding ductility ratio (μ_{sust}) at maximum sustained displacement

Specimen	δ_{yn} (mm)	δ_{sust} (mm)	μ_{sust}
CFP-0	7.14	32.07	4.5
EC-0	7.14	26.3	3.6
CFP-400	6.8	23.27	3.4
EC-400	6.8	23.27	3.4
CFP-500	6.4	23	3.6
CFP-675	5.1	20.06	4

the critical lengths of the specimens, when also subjected to the action of an axial load, does not lead to an improvement in structural behaviour other than, perhaps, a slight reduction in the rate of loss of load-carrying capacity during the last cycles of loading to the displacement that leads to failure of both types of specimens. It should be noted that, for the case of $N = 400$ kN, both specimens fail due to out of plane displacements induced by asymmetric cracking (with respect to the longitudinal plane of symmetry of the specimens) combined, perhaps, with an unintended out of plane eccentricity of the applied axial load.

Figure 7.6a–d indicate that all specimens designed in accordance with the proposed method sustained maximum displacements corresponding to values of the ductility ratio of the order of 4; these values are significantly larger than the value of 2 specified by EC8 for high ductility determinate RC structural members [see clauses 5.2.2.1 and 5.2.3.4(3) and Table 5.1 of EC8]. It appears, therefore, that, when $N \neq 0$, the likelihood of P-δ effects due to out of plane displacements (induced, as discussed earlier, by asymmetric cracking of the specimens combined with a possible unintended out of plane eccentricity of N) does not have any practical effect on safeguarding the code specified limiting values for ductility.

A comparison of the calculated values of load-carrying capacity of the specimens tested under cyclic loading with their experimentally-established counterparts (see Table 7.1) indicates that the calculated values consistently underestimate the experimentally-established ones by about 8 %. Both values of load-carrying capacity correspond to a flexural mode of failure, with the calculation of the flexural capacity M_f being based on the assumption that the steel stress after yielding remains constant and equal to the yield stress f_y. The validity of this assumption is subsequently verified through a comparison of the calculated values of the steel strains with the maximum value of the yield plateau of the experimentally-established stress–strain curves of the steel used, which shows that the former value is smaller than the latter. It appears, therefore, that ignoring the steel hardening properties of the steel is not the main cause of the above deviation, as usually suggested (EC8) and discussed in Sect. 3.3.1.

Finally, from Fig. 7.7 it can be seen that all specimens exhibited a flexural mode of failure. It should be noted, however, that the spalling of concrete at the side face of the specimens subjected to an axial compressive force is due to the out of plane bending (discussed earlier) which precedes the loss of load-carrying capacity. On the other hand, for the specimens subjected only to cyclic loading, the flexural mode of failure is also characterised by extensive bond failure along both the top and bottom longitudinal reinforcement.

7.2.5 Concluding Remarks

Designing in accordance with the proposed method leads to significant savings in stirrup reinforcement without compromising the code requirements for structural performance. In fact, the code specified stirrup reinforcement can be impractical as it leads to reinforcement congestion in the critical lengths when the axial force exceeds a certain threshold, which, for the specimens investigated, is around $N = 0.2 N_u$.

More specifically, for values of the axial force N up to $0.15 N_u$, structural behaviour is found independent of the method of design adopted, whereas all specimens designed in accordance with the proposed method exhibit a flexural mode of failure and ductility significantly larger than the code specified value.

7.3 Structural Walls

Current code (e.g. EC2 [9], EC8 [5]) provisions for the earthquake-resistant design of reinforced-concrete (RC) structural walls (SW) specify reinforcement arrangements comprising two parts: One part forming "concealed column (CC)" elements (usually extending between the ground and first floor levels of buildings) along the two vertical edges of the walls; the other consisting of a set of grids of uniformly distributed vertical and horizontal bars, within the wall web, arranged in parallel to the wall large side faces. The CC elements are intended to impart to the walls the code specified ductility, whereas the wall web is designed against the occurrence of "shear" failure, before the wall flexural capacity is exhausted. The specified ductility is considered to be achieved by confining concrete within the CC elements through the use of a dense stirrup arrangement—thus increasing both the strength and the strain capacity of the material; on the other hand, shear failure is mainly prevented by providing horizontal web reinforcement capable of sustaining either the whole or the portion of the shear force in excess of that that can be sustained by concrete alone. It is also important to add that the calculation of the wall flexural capacity allows for the contribution of all vertical reinforcement, within both the CC elements and the web.

The above design procedure, however, has a significant drawback: the dense spacing of the stirrups often results in reinforcement congestion within the CC elements and this may cause difficulties in concreting and, possibly, incomplete compaction of the concrete. Recently published work has shown that reinforcement congestion within the CC elements can be prevented without lowering the load-carrying capacity and the ductility (current code procedures are considered to safeguard against) by designing the walls in accordance with the proposed method [1, 10]. The work is experimental in nature and based on a comparison of the results obtained from tests on walls designed in accordance with current code provisions for the earthquake-resistant design of RC structures with those obtained from tests on walls with the same geometry and flexural reinforcement, but with transverse reinforcement designed in accordance with the proposed method. The work outlined in the following is extracted from Ref. [1] where full details may be found regarding the case of walls of type II behaviour. The case of walls of type III behaviour forms the subject of Ref. [10] where the conclusions drawn are in agreement with those drawn for the case of type II walls discussed in what follows.

7.3.1 Experimental Details

The structural walls investigated are designated by using a three part name, with the first part indicating the method of design (EC or CFP), the second the diameter (10 or 12 mm) of the longitudinal (flexural) bars, and the third the type of loading (monotonic M or cyclic C). The total number of walls tested is eight, and their design details are shown in Figs. 7.8 and 7.9 for the walls designed in accordance with the code provisions and the CFP method, respectively. The figures indicate that all walls have a length $l = 700$ mm, a height $h = 1,700$ mm and width $b = 100$ mm. The longitudinal reinforcement comprises eight pairs of steel bars with a 10 or 12 mm diameter, 100 mm centre-to-centre spacing, and a distance of a bar's centre from the wall face closest to it equal to 25 mm.

In contrast with the longitudinal reinforcement, the amount and arrangement of the horizontal reinforcement placed in the walls depends on the method of design employed. For all walls designed in compliance with the EC2 and EC8 provisions, the horizontal reinforcement comprises 8 mm diameter stirrups at a

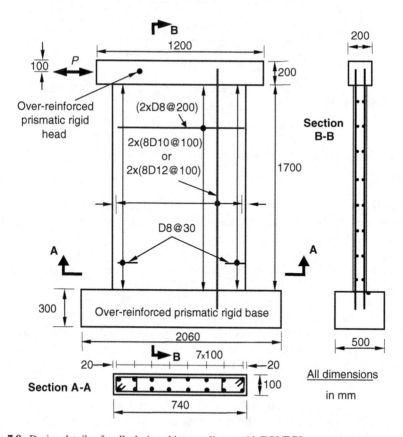

Fig. 7.8 Design details of walls designed in compliance with EC2/EC8

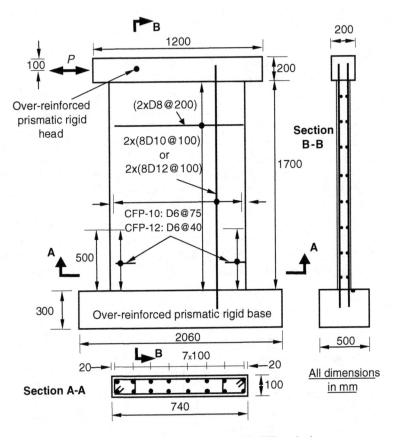

Fig. 7.9 Design details of walls designed in compliance with CFP method

centre-to-centre spacing of 30 mm within the CC elements extending throughout the specimen height,. On the other hand, for the specimens designed in accordance with the CFP method, the stirrups have a 6 mm diameter and a centre-to-centre spacing of 75 or 40 mm for the walls with a 10 or 12 mm longitudinal bar diameter, respectively, and extend to a distance from the wall lower end equal to about 500 mm. For all walls, the web horizontal reinforcement comprises a pair of 8 mm diameter bars at a centre-to-centre spacing of 200 mm. The strength characteristics of both concrete and steel reinforcement are given in Table 7.4.

The walls are subjected to two types of horizontal loading applied along the horizontal axis of symmetry of the rigid prismatic element monolithically connected to the walls at their top face (see Figs. 7.8 and 7.9):

(a) Static load monotonically increasing to failure;
(b) Static cyclic loading applied in the form of horizontal displacements varying between extreme predefined values, initially equal to ±10 mm and increasing in steps of 10 mm thereafter until failure of the specimens.

Table 7.4 Strength characteristics of concrete and steel reinforcement

Specimens	f_c	D6		D8		D10		D12	
		f_y	f_u	f_y	f_u	f_y	f_u	f_y	f_u
	(MPa)	(MPa)	(MPa)	(MPa)	(MPa)	(MPa)	(MPa)	(MPa)	(MPa)
CFP-10-M	25	395	482	563	667	555	655	–	–
EC-10-M	25	–	–	563	731	555	655	–	–
CFP-10-C	35	395	482	563	667	621	697	–	–
EC-10-C	32	–	–	563	667	555	655	–	–
CFP-12-M	29	395	482	563	667	–	–	554	678
EC-12-M	29	–	–	563	667	–	–	554	661
CFP-12-C	25	574	634	563	731	–	–	554	661
EC-12-C	25	–	–	563	667	–	–	554	661

Three load cycles are carried out for each of the above predefined values with a displacement rate of 0.25 mm/s.

Full details of the experimental set up used for the tests are provided elsewhere [1].

7.3.2 Design

The walls are designed so that their load-carrying capacity is reached when their base cross-section attains its flexural capacity, the latter condition being referred to henceforth as *plastic-hinge* formation. Flexural capacity, denoted as M_f, is calculated as described in Sect. 3.3.1, allowing for the contribution of all vertical reinforcement, both within the CC elements and within the web and assuming material safety factors equal to 1. Using M_f, the wall load-carrying capacity P_f (and, hence, the corresponding shear force $V_f = P_f$) is easily calculated from static equilibrium. The values of M_f and $V_f = P_f$ for each of the specimens tested are given in Table 7.5 together with the experimentally-established values of the load-carrying capacity P_{exp}. The table also includes the values of bending moment M_y and load P_y which correspond at the yielding of the flexural reinforcement closest to the tensile face of the walls; the latter values are used in order to assess the ductility ratio of the specimens tested.

As mentioned earlier, the horizontal reinforcement of the walls is designed either in compliance with the clauses of EC2 and EC8 for the design of earthquake-resistant RC structures or in accordance with the CFP method. It is interesting to note in Fig. 7.8 the densely spaced stirrups confining the CC elements within the "critical regions" (extending throughout the wall height) specified by the Codes. Such spacing [resulting from expressions 5.20 in EC8 (clause 5.4.3.2.2)] is considered to safeguard ductile wall behaviour. As discussed in Sect. 4.3.1, in contrast with the code reasoning behind the calculation of the stirrups within the CC elements, the CFP method specifies a significantly smaller amount of such reinforcement only for case of walls of type II behaviour, not for providing confinement to concrete, but in order to sustain the transverse tensile

Table 7.5 Calculated values of bending moment M_y and corresponding force P_y at yield, flexural capacity M_f and corresponding load-carrying capacity P_f, experimentally-established values of load-carrying capacity P_{exp} and ductility ratio μ_{exp}

	My (kNm)	Py (kN)	Mf (kNm)	Pf (kN)	Pexp (kN)	μexp
CFP-10-M	133	74	241	134	150	3.9
EC-10-M	133	74	241	134	148	3.9
CFP-10-C	135	75	241	134	141(+)/112(−)	3.3
EC-10-C	150	83	257	143	136(+)/114(−)	3.4
CFP-12-M	185	103	300	167	188	3.6
EC-12-M	185	103	300	167	182	3.6
CFP-12-C	184	102	300	167	171(+)/154(−)	2.25
EC-12-C	184	102	300	167	166(+)/150(−)	2.25

stresses developing within the compressive zone as a result of stress redistribution due to the loss of bond between concrete and the flexural reinforcement at the base of the walls. (It should be noted that the walls in Figs. 7.8 and 7.9 are indeed of type II behaviour, since, for shear span $a_v = 1{,}800$ mm and cross-sectional depth (i.e. distance of the resultant of the forces sustained by the flexural tension reinforcement from the extreme compressive fibre) $d \approx 450$ mm, $a_v/d \approx 4 > 2.5$. On the other hand, as discussed in Sect. 4.3.2 for the case of type III behaviour, the CFP method does not recognize the need for additional transverse reinforcement within the critical regions.

As regards the horizontal web reinforcement, that designed in compliance with the code requirements (see clauses 6.2 and 9.6 in EC2) is considered to prevent shear failure of the walls before their flexural capacity is exhausted. This reasoning is also in conflict with that underlying the design of the horizontal reinforcement within the wall web in accordance with the CFP method: in the latter case, as regards type II behaviour, horizontal reinforcement is designed so as to sustain the tensile force developing in the region of change in the direction of the path of the compressive force developing due to the bending action (see Sect. 4.3.1); as regards type III behaviour, such reinforcement is designed to sustain the horizontal force required to develop in order to produce additional flexural resistance, which, when added to the bending moment corresponding to the wall's load-carrying capacity in the absence of horizontal reinforcement, the resulting bending moment equals the flexural capacity of the wall's cross section (see Sect. 4.3.2).

7.3.3 Results

The main results of the work are given in Figs. 7.10, 7.11, 7.12, 7.13, 7.14, 7.15. Figures 7.10 and 7.11 show the load–deflection curves obtained under statically-applied monotonic loading, whereas the load–deflection curves obtained under statically-applied cyclic loading are shown in Figs. 7.12 and 7.13. Finally, typical modes of failure of the walls under monotonic and cyclic loading are shown in Figs. 7.14 and 7.15.

Fig. 7.10 Load-deflection curves for walls CFP-10-M and EC-10-M under monotonic loading

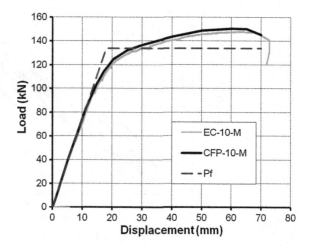

Fig. 7.11 Load-deflection curves for walls CFP-12-M and EC-12-M under monotonic loading

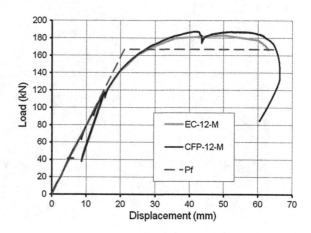

Fig. 7.12 Load-deflection curves for walls CFP-10-C and EC-10-C under cyclic loading

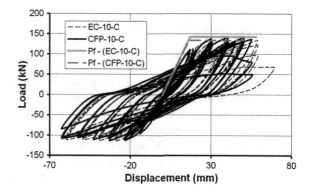

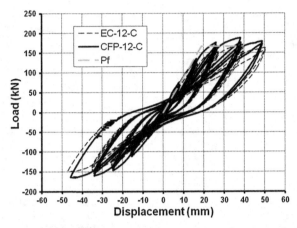

Fig. 7.13 Load-deflection curves for walls CFP-12-C and EC-12-C under cyclic loading

Fig. 7.14 Typical failure mode of walls under monotonic loading with the location of failure indicated in a magnified form at the *bottom*

7.3.4 Discussion of Results

Monotonic loading

As indicated in Figs. 7.10 and 7.11, in spite of the differences in reinforcement arrangement, the two types of walls investigated exhibit similar behaviour under monotonic loading: Walls CFP-10-M and CFP-12-M exhibit a slightly larger load-carrying capacity and stiffness than those of walls EC-10-M and EC-12-M, respectively, with all walls exhibiting similar values of maximum displacement. Also, all walls exhibit a similar mode of failure in that the loss of load-carrying capacity is preceded by failure of the compressive zone at the wall base (see Fig. 7.14). Such

Fig. 7.15 Typical failure mode of walls under cyclic loading with the location of failure indicated in a magnified form at the *bottom*

behaviour clearly demonstrates that, under this type of loading, any amount of reinforcement larger than that specified by the CFP method is essentially ineffective.

Figures 7.10 and 7.11 also show the location of the point of nominal yield used for assessing the specimen ductility ratio. The location of this point is determined as discussed in Sect. 7.2.3 for the case of beam-column elements.

It is evident that all monotonically loaded specimens exhibit ductile behaviour. In fact, Table 7.5 indicates that the average value of the ductility ratio (μ_{exp}) of the specimens is nearly 4.

Cyclic loading

Figures 7.12 and 7.13 indicate that, as for the case of the monotonic loading, both types of walls exhibit similar behaviour, with walls CFP-10-C and CFP-12-C being characterised by a slightly larger load-carrying capacity in spite of the significantly less amount of stirrup reinforcement within the CC elements. On the other hand, as indicated in Figs. 7.14 and 7.15, the failure mode (crushing of the compressive zone at the base of the walls) is found to be independent of the loading regime imposed. It is interesting to note, however, that this mode of failure marks the start of an abrupt loss of load-carrying capacity in all cases investigated.

7.3.5 Concluding Remarks

Designing in accordance with the CFP method leads to significant savings in horizontal reinforcement without compromising the code performance requirements.

More specifically, the amount of stirrup reinforcement specified by the CFP method is significantly lower than that specified by current codes; moreover, such reinforcement is placed within a portion of the concealed-column elements extending to just over one-third of the wall height, as compared with the full element height recommended by the codes. In contrast with the case of the stirrups, the amount of horizontal web reinforcement specified by the CFP method is similar to the code specified amount for the structural walls investigated.

7.4 Beam-Column Joints

The results presented in the following provide an indication of the behaviour, under simulated seismic loading, of beam-column joints designed in accordance with the proposed method. They have been extracted from Ref. [4], where full details of the work can be found, and their discussion complements the discussion in Sect. 6.4.3.

7.4.1 Experimental Details

The structural forms investigated are eight full-size beam-column joint sub-assemblages schematically represented in Fig. 7.16. The linear elements (beam and columns) of these sub-assemblages represent the portion of the elements between the joint and the nearest point of contraflexure (point of zero bending moment). As indicated in the figure, the specimens are pin-supported at the end of the lower column and simply-supported at the beam end; they are subjected to horizontal displacement of the free end of the upper column, with the distance between the displacement point and the pin support being 3.00 m symmetrically

Fig. 7.16 Schematic representation of specimens investigated

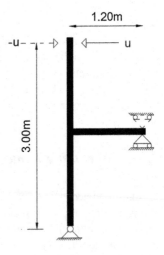

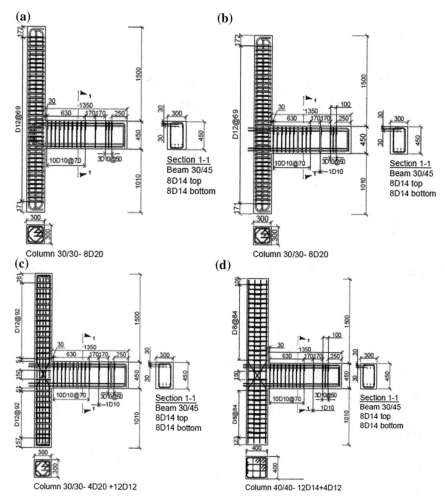

Fig. 7.17 Design details of specimens (a) S1, (b) S2 and S2′, (c) S5, (d) S6, (e) S9, (f) S10, (g) S11 and (h) steel plates of two rows of longitudinal beam reinforcement (dimensions in mm)

extending about the longitudinal axis of the beam, whereas the distance between the simple support and the longitudinal axis of the column is 1.2 m. Full details of the experimental set up and the measuring techniques employed are provided in Ref. [4].

The design details of four of the specimens tested, denoted as S1, S2, S2′, and S5, are shown in Fig. 7.17a–c. The mean value of the cylinder compressive strength of concrete at the time of testing is $f_c = 35$ MPa, whereas the stress–strain characteristics of the steel reinforcement used are indicated in Table 7.6. For all specimens, the beam and column elements are designed in compliance with the provisions of EC2 and EC8 for high ductility members

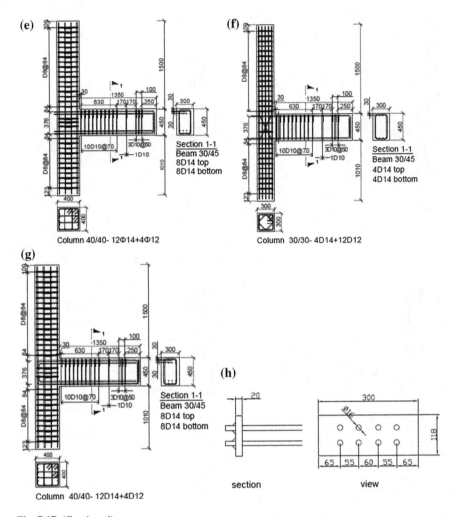

(e)

Section 1-1
Beam 30/45
8D14 top
8D14 bottom

Column 40/40- 12Φ14+4Φ12

(f)

Section 1-1
Beam 30/45
4D14 top
4D14 bottom

Column 30/30- 4D14+12D12

(g)

Section 1-1
Beam 30/45
8D14 top
8D14 bottom

Column 40/40- 12D14+4D12

(h)

section   view

Fig. 7.17 (Continued)

Table 7.6 Stress-strain properties of steel reinforcement used for the specimens

Diameter	Yield stress f_y (MPa)	Strength f_u (MPa)	Normalized stress–strain curve for steel bars
Φ8	633	729	
Φ10	587	714	
Φ12	571	686	
Φ14	563	678	
Φ20	560	670	

(HDC). The joint transverse reinforcement in specimens S1, S2 and S2' is also designed in accordance with the European Codes (EC2, EC8) by extending the column stirrups through the joint, with the beams' longitudinal bars of specimen S1 being anchored as specified by the codes, whereas those of specimens S2, S2' and S5 are welded on to steel plates placed at the free face of the joint. On the other hand, for specimen S5, the joint is reinforced with diagonal reinforcement by extending the column longitudinal bars (between the corner bars) diagonally within the joint as indicated in Fig. 7.17c, rather than by placing additional reinforcement as opted elsewhere [6.28]; in this case, the amount of joint stirrup reinforcement is reduced to 25 % of that for specimens S1, S2, S2', with the use of inclined reinforcement not adding to reinforcement congestion in the joint as it is formed from the column longitudinal bars.

The design details of the remaining four specimens denoted as S6, S9, S10, and S11 are indicated in Fig. 7.17d–g. The joints of all specimens are designed as described in Sect. 6.4.4. The resulting amount of reinforcement is placed either in the form of inclined bars, evenly distributed across the joint width by extending the column longitudinal bars (between the corner bars) diagonally into the joint, or in the form of horizontal stirrups placed between the tensile and compression longitudinal bars of the beam.

For specimens S5, S9 and S10, the beam longitudinal bars are welded onto steel plates as for specimens S2, S2' and S5 (see Fig. 7.17h), whereas for specimen S11 the beam longitudinal reinforcement is anchored within the joint as for specimen S1. The joint shear reinforcement of specimens S9 and S11 consists of three four-legged 10 mm diameter stirrups, whereas specimens S6 and S10 are reinforced with diagonal reinforcement as for specimen S5, with the joint stirrup reinforcement of specimen S6 being reduced to 42 % of that of specimens S9 and S11.

With the exception of specimen S2' which is subjected to monotonic loading, all other specimens are subjected to cyclic loading applied in the form of horizontal displacements varying between extreme predefined values equal to $0.5\delta_y$, $1.0\delta_y$, $1.5\delta_y$, $2.0\delta_y$, $2.5\delta_y$, and $3.0\delta_y$, where δ_y is the horizontal displacement of the load point at yield of the specimen. Three load cycles are carried out for each of the above predefined values with a loading rate of 0.25 mm/s.

Displacement δ_y is assessed from the load–displacement curve established for specimen S2' under monotonic loading by assuming that yield occurs when the applied load attains a value equal to $0.7P_u$ (where P_u is the experimentally-established load-carrying capacity) and it is found to be equal to approximately 50 mm [5]. The displacement δ_y for specimen S10 is assumed to be half that of specimen S2' (i.e. 25 mm), since its beam longitudinal reinforcement is reduced to half.

The test is terminated when the residual load-carrying capacity of the specimen reduced below 85 % P_u (EC8) within the first cycle of cyclic-load sequence at a predefined displacement value (where P_u is the specimen load-carrying capacity).

7.4.2 Results

Full results of the tests may be found in Ref. [4]. Herein, the results presented
are load–displacement curves and crack patterns of the specimens tested at various
stages of the induced displacement. Both loads and displacements of the cyclic
loading tests are expressed in a non-dimensional form; they are normalized with

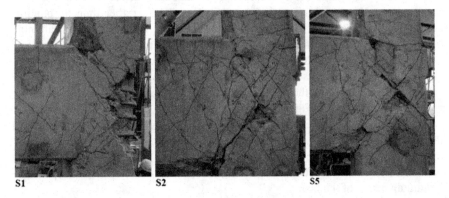

S1   S2   S5

Fig. 7.18 Crack pattern of specimens S1, S2 and S5 at failure

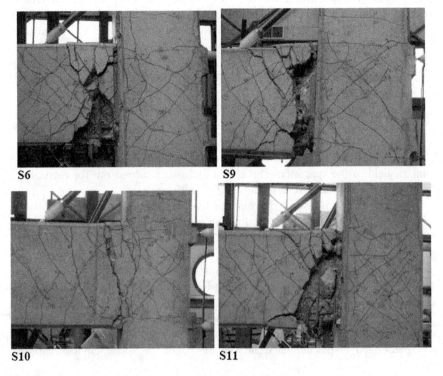

S6   S9

S10   S11

Fig. 7.19 Crack pattern of specimens S6, S9, S10 and S11 at failure

respect to the load-carrying capacity (P_{fb}) and displacement at yield (δ_y), respectively, the latter being defined in a preceding section and the former is the calculated value corresponding to the flexural capacity of the beam.

The failure modes of all specimens at the end of the test are shown in Figs. 7.18 and 7.19, whereas Figs. 7.20 and 7.21 show the crack patterns of two typical specimens with end plates, specimens S2 and S6, respectively, at the

Fig. 7.20 Crack pattern of specimens S2 at the end of the cyclic stages at maximum displacements equal to (**a**) 25 mm, (**b**) 50 mm, (**c**) 75 mm, (**d**) 100 mm, (**e**) 125 mm, and (**f**) 150 mm

Fig. 7.21 Crack pattern of specimens S6 at the end of the cyclic stages at maximum displacements equal to (**a**) 25 mm, (**b**) 50 mm, (**c**) 75 mm, (**d**) 100 mm, (**e**) 125 mm, and (**f**) 150 mm

end of each of the cyclic stages of the loading regime. The full non-dimensional load–displacement curves for all specimens tested under cyclic loading are presented in Figs. 7.22 and 7.23.

Finally, Table 7.7 provides the calculated values of the load (P_{fb}) corresponding to the formation of plastic hinges in the beam, as well as to the calculated values of the acting shear force (V_{jh}) and shear capacity of the joint according to EC8 (V_{Rjh}) and the proposed method (F_{Rjh}), with the safety factors being taken equal to 1. The table also includes the experimentally-established values of load-carrying capacity for all specimens tested together with an indication of their experimentally established and predicted modes of failure.

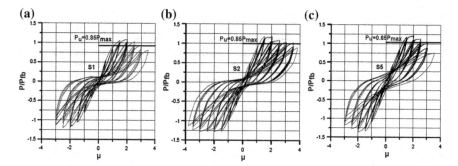

Fig. 7.22 Non-dimensional load–displacement curves for specimens: **a** S1, **b** S2, and **c** S5 under cyclic loading

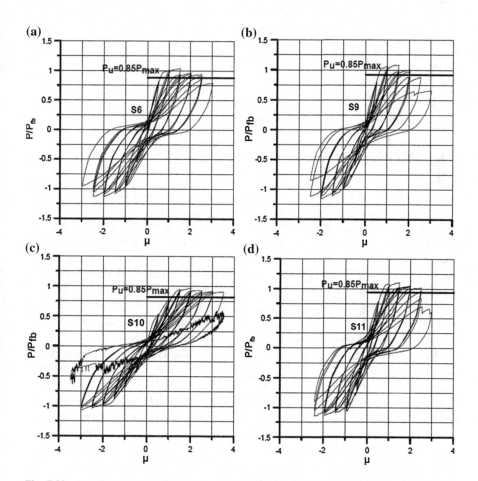

Fig. 7.23 Non-dimensional load–displacement curves for specimens: **a** S6, **b** S9, **c** S10, and **d** S11 under cyclic loading

Table 7.7 Calculated and experimental load-carrying capacity of specimens tested

Specimen	P_{fb} (kN)	V_{fjh} (kN)	EC8		Proposed method		P_{EXP} (kN)	Experimental failure mode
			V_{Rjh} (kN)	FM	F_{Rjh} (kN)	FM		
S1	116	716	1,040	BF	539	JF	99–107	JF
S2	116	716	1,040	BF	539	JF	106–115	JF
S3	116	716	1,040	BF	539	JF	95–115	JF
S4	116	716	1,040	BF	539	JF	102–113	JF
S5	115	717	1,040	BF	538	JF	109–122	JF
S6	110	722	1,965	BF	1,291	BF	108–118	BF
S7	111	721	1,965	BF	1,291	BF	108–127	BF
S8	111	721	1,965	BF	1,291	BF	109–123	BF
S9	111	721	1,965	BF	1,291	BF	113–120	BF
S10	61	355	1,040	BF	532	BF	59–64	BF
S11	111	722	1,965	BF	1,291	BF	115–119	BF

P_{fb} calculated load-carrying capacity corresponding to flexural capacity of the beam
V_{fjh} calculated joint shear force corresponding to flexural capacity of the beam
V_{Rjh} calculated joint shear capacity according to EC8
F_{Rjh} calculated joint shear capacity according to proposed method
FM mode of failure
BF formation of plastic hinge in the beam
JF extensive cracking in the joint
P_{EXP} experimentally established load-carrying capacity of specimens
For all calculated values safety factors are set equal to 1

7.4.3 Discussion of Results

A major requirement of the seismic design of RC structures concerning beam–column joints is to minimize the likelihood of shear failure of the joint before the formation of a plastic hinge in the beam or column regions adjacent to the joint, with plastic hinge formation in the beam preceding plastic hinge formation in the column. However, for specimens S1, S2 and S5, the wide flexural cracks developed in the beam does not appear to have prevented significant diagonal cracking of the joint (see Fig. 7.18), the condition of which is found to deteriorate with each load cycle (see Fig. 7.20). On the other hand, the failure mode of specimens S6, S9, S10 and S11 involves the formation of a plastic hinge in the beam, as indicated by the formation of wide flexural cracks and spalling of the concrete cover to both the longitudinal and the transverse reinforcement (see Fig. 7.19), with the joints suffering significantly less cracking which, once it forms, remains essentially unchanged throughout the duration of the test (see Fig. 7.21).

From the results included in Table 7.7 for specimens S1, S2 and S5, it appears that EC8 overestimates the shear capacity of the joint, since the specimens fail before, rather than after, the formation of a plastic hinge in the beam, the latter corresponding to a value of 716 kN for the acting shear force which is significantly smaller than the code predicted value of 1,040 kN for the joint shear capacity. On the other hand, the proposed method correctly predicts failure of the joint to occur before the formation of a plastic hinge in the beam, since the value of

539 kN predicted for the joint shear capacity is smaller than the acting shear force of 716 kN corresponding to the formation of a plastic hinge in the beam.

As regards specimens S6, S9 and S11, Table 7.7 indicates that the method adopted for calculating the joint shear strength achieves its purpose, since failure of the joints is prevented in all cases. It may also be of interest to note in the table that the calculated values of the specimens' load-carrying capacities corresponding to the formation of a plastic hinge in the beam only slightly underestimate the experimentally established values by a value ranging between 1 and 5 %.

It is interesting to note that for the joints of specimens S5 and S10, which have the same geometrical characteristics and arrangement of reinforcement (see Fig. 7.17c–f), the joint shear strength in accordance with the Eurocodes is about 1,040 kN, whereas the value accessed through the use of the proposed method is approximately equal to 538 kN, i.e. a little more than half the code value. However, as indicated in Table 7.7, the experimentally established values of the load-carrying capacity of the specimens designed in accordance with the code provisions are smaller than the values corresponding to the formation of a plastic hinge in the beam, whereas the value (717 kN) of the acting shear forces corresponding to the formation of a plastic hinge in the beam is larger than the value (539 kN) predicted by the proposed method. This explains the inability of the joints of specimens S1, S2 and S5 with a 300 mm width to sustain the forces induced to them by the beam and column elements of specimens. It appears, therefore, that failure of the joint could be prevented either by increasing its width or by reducing the forces transferred to the joint by the beam and column elements through a reduction of the longitudinal reinforcement.

In order to prove this point the following modifications are implemented to specimens S6, S9 and S11: as discussed earlier, the column and the joint width is increased to 400 mm, while for specimen S10 the beam longitudinal reinforcement is reduced to half that of specimens S1, S2 and S5. In fact, from Fig. 7.19 it can be seen that all four specimens satisfy the performance requirements under cyclic loading; the formation of the plastic hinge in the region of the beam adjacent to the joint occurs before any significant distress of the joint.

Figures 7.22 and 7.23 show the load–displacement curves under cyclic loading for specimens S1, S2 and S5 and S6, S9, S10 and S11, respectively. The progressive cracking of specimens S1, S2 and S5 with each load cycle leads to a reduction in stiffness, with the cracking of the joint being the underlying cause of the "pinching" effect characterising the load–displacement curves of the specimens, whereas the reduction in stiffness of specimens S6, S9, S10 and S11 appears to be due to the cracking of the beam.

From a more detailed comparison of the load–displacement curves presented in Ref. [4], specimens S2, S5, S6, S9, and S10, whose beam longitudinal reinforcement is welded onto steel plates at the free side of the joint, are found to behave better than specimen S1 and S11 whose beam longitudinal bars are bent near the free end of the joint in the manner specified by the EC2 and EC8. The beneficial effect of the end plates is apparent also in the cases in which the code performance requirements are not achieved [4]. This beneficial effect has been attributed to the more uniform distribution of the compressive force achieved through the use of an end plate, which enables the development of a wider compressive strut in the joint, thus leading to an increase in the joint load-carrying capacity.

Moreover, as shown in Fig. 7.17c–d and f, the transverse reinforcement of specimens S5, S6 and S10 is aligned diagonally within the joint and consists of six 12 mm diameter bars for specimens S5 and S10 and of eight 14 mm diameter bars for specimen S6. For specimens S5, the stirrup joint reinforcement is 75 % less than that of specimen S2, whereas, for specimen S6 is 80 % less than that of specimen S9. It is apparent from Fig. 7.22b–c that the hysteretic loops of specimen S5 are characterised by a significantly less pronounced pinching than specimen S2, although the later suffers a smaller loss of load-carrying capacity in the last loop of the cyclic loads at a normalised displacement equal to 3.5. In fact, such an improvement in behaviour occurs in spite of the significant reduction of the stirrup reinforcement in the joint to nearly 25 % the amount specified by EC2 and EC8.

It appears from the above that the amount of horizontal stirrups specified by EC8 for designing earthquake-resistant joints, not only is found incapable to fulfil its purpose, but also results in steel congestion in the joint which causes many difficulties in placing and compacting concrete; on the other hand, for the specimens investigated, the proposed method for calculating joint shear reinforcement results in 73 % less joint stirrup reinforcement and yet achieves the code requirements for structural performance. Moreover, the inclined bars, which are not additionally placed in the joint, but form part of the column longitudinal reinforcement, are more effective than the horizontal stirrups, since they are placed at nearly right angles to the direction of cracking.

Finally, it should be reminded that the verification of the proposed method for designing beam-column joints has also been based on a comparative study of the predicted behaviour with its counterpart established from test results reported in the literature; this comparative study forms the subject of Sect. 6.4.5 of the preceding chapter.

7.4.4 Concluding Remarks

The provisions of EC2 and EC8 are not found capable of preventing significant diagonal cracking of the joint before the formation of a plastic hinge in the beam element. In fact, specimen S1, which is designed in compliance with the codes, is found to suffer significant cracking in the joint; this leads to a progressive loss of joint stiffness before the beam element suffering any apparent loss in load-carrying capacity.

The code recommended methods are also found unable to provide a realistic assessment of the joint shear capacity which appears to be overestimated. This may have been one of the main causes of the significant cracking and loss of stiffness suffered by the joints of the specimens discussed.

The amount of horizontal stirrups specified by EC8 for designing earthquake-resistant joints results in steel congestion; the latter causes many difficulties in placing and compacting concrete. And yet, as discussed earlier, the structural performance code requirements are not achieved.

The method proposed for calculating the joint shear strength is found capable of giving a realistic assessment of the joint strength. The specimens designed according to the proposed method fulfil the code requirements for structural performance. Their failure mode involves the formation of a plastic hinge in the beam, as indicated by the formation of wide flexural cracks and spalling of the concrete cover to both longitudinal and transverse reinforcement, with the joints suffering significantly less cracking which, once it forms, remains essentially unchanged throughout the duration of the test.

The use of steel plates for anchoring the beam's longitudinal reinforcement is found to improve specimen behaviour as regards both load-carrying capacity and stiffness.

The proposed method for calculating joint shear reinforcement results in up to 73 % less joint stirrup reinforcement than the code specified amount, without any compromise in the code requirements for structural performance.

For the specimens with steel plates for anchoring the beam's longitudinal reinforcement, the use of inclined bars instead of horizontal stirrups in the joint, leads to an improvement of structural behaviour. Such an improvement is achieved in spite of the use of much less stirrups reinforcement.

7.5 Points of Contraflexure

As discussed in Sect. 1.5, a common type of damage of RC buildings is the unexpected brittle failure suffered by columns at the location of the points of contraflexure during the 1997 earthquake that occurred in Athens [2]. Similar types of failure, which cannot be attributed to either non-compliance with code provisions or defective work [2], are reported to have been suffered by the vertical, rather than the horizontal, elements of RC structures, not only during severe earthquakes that have occurred throughout the world over the last 15 years [11, 12], but, also, under static loading conditions [13].

The above types of failure are not taken into consideration by the methods adopted by current codes of practice for the design of RC structures, invariably based on the truss analogy (TA) [14]. As discussed in Sect. 6.2.1, the proposed method, in contrast with design based on the TA, recognizes the location of the point of contraflexure as one of the potential locations of weakness of linear structural members. In fact, the causes of this type of failure have been investigated by experiment by reproducing them under controlled laboratory conditions [15–18]. The investigation has been based on a comparative study of current code predictions with the results obtained from test on linear structural concrete members subjected to the combined action of axial and transverse forces. The specimens have been designed, on the one hand, from first principles using the TA (as applied by ACI and EC2, EC8) and, on the other hand, in accordance with the proposed method. The indicative results presented in the following are extracted from Ref. [18], where full details can be found.

7.5.1 Experimental Details

The specimens investigated are simply supported two-span linear elements, with a rectangular cross section, subjected to the combined action of an axial-compressive concentric force N and a transverse force P which, as shown in Fig. 7.24, is applied at mid length of the longer span. The figure also includes the bending moment M and shear force V diagrams corresponding to the formation of one and two plastic hinges at the locations where the element cross section reaches its flexural capacity M_f. The first plastic hinge occurs at the location of the transverse point load, while the second occurs at the location of the internal support. The formation of the first plastic hinge may be considered as a lower-bound limit for load-carrying capacity, while the formation of the second plastic hinge transforms the element into a mechanism, which may be considered as an upper bound limit for structural collapse. Moreover, it may be noted from the internal-force diagrams of Fig. 7.24 that the portion of the elements between the internal support and the transverse point load may be viewed as an approximately 1:3 scale physical model of the portion of a column between two successive stories of a building.

The specimens are subjected to sequential loading: the axial force N is applied first. It increases to a predefined value equal to $N \approx 0.2N_u$ ($N_u = f_c bh$, f_c is the

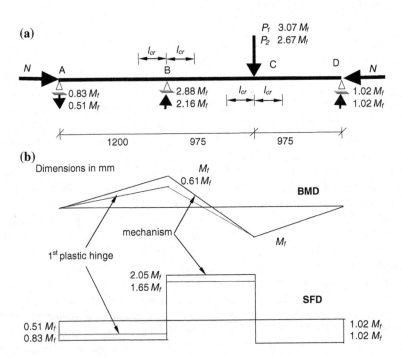

Fig. 7.24 Structural forms investigated: **a** applied forces; and **b** bending moment M and shear force V diagrams corresponding to formation of one and two plastic hinges (ph)

uniaxial cylinder compressive strength of concrete, and b, h are the cross-sectional dimensions of the specimen), where it is maintained constant during the subsequent application of the transverse load P. The latter force (applied at the middle of the larger span) increases to failure either monotonically, or in a cyclic manner inducing progressively increasing displacements in opposite directions as shown in Fig. 7.2.

As discussed in the preceding section, the specimens are designed from first principles by using methods based on two contrasting concepts: the TA, as applied by ACI and EC2/EC8, and the proposed method. The physical models underlying the design methods are shown in Fig. 7.25. It is interesting to note in the figure that, unlike TA, the CFP concept recognizes the locations of the structural element where the development of transverse tension is most likely to cause non-flexural types of failure. In fact, for the structural elements investigated, the latter concept predicts the region of location 3 [the location of the point of contraflexure (see Fig. 7.25b)] as the region most likely to fail in transverse tension; as a result the stirrup spacing is significantly denser than that code specified value in this region (see Fig. 7.26). On the other hand, both ACI and EC2/EC8 specify denser stirrup spacing in regions where a large shear force combines with a large bending moment. Such regions, referred to in codes as critical lengths, are marked with l_{cr} in Fig. 7.24, with $l_{cr} = 400$ mm (see Fig. 7.26a and b). With the exception of the transverse reinforcement arrangement, shown in Fig. 7.26, the design details of the specimens are similar in all cases. The specimens have a 200 mm square cross section and longitudinal reinforcement comprising four 16 mm diameter high yield deformed bars (with yield stress $f_y = 540$ MPa), symmetrically placed at the corners of the cross section. The distance of the centroid of the longitudinal bars from the top and bottom faces closer to them is approximately 15 mm. Mild-steel plain

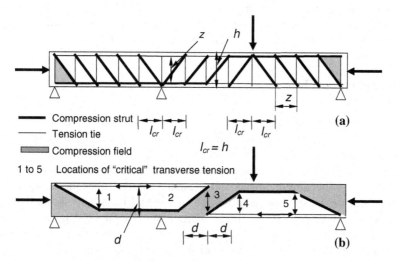

Fig. 7.25 Models underlying methods used for designing structural forms tested: **a** truss analogy (*TA*); and **b** compressive force path model (*CFP*)

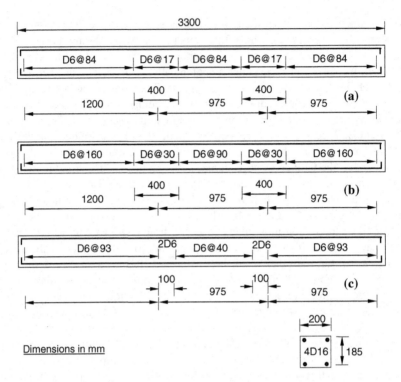

Fig. 7.26 Design details of specimens with 14 mm diameter longitudinal bars: **a** design to TA (ACI); **b** design to TA (EC); **c** design to CFP

bars with a 6 mm diameter with yield stress $f_y = 360$ MPa are used for the stirrups. The ready-mix concrete used has an uniaxial cylinder compressive strength $f_c \approx 40$ MPa at the time of testing (nearly 60 days after casting).

7.5.2 Results of Tests

Typical test results are presented in Tables 7.8, 7.9, 7.10, 7.11 and Figs. 7.27, 7.28, 7.29, 7.30. Table 7.8 shows the experimental and calculated values of the transverse load and corresponding displacement at various load stages for all specimens considered. Table 7.9 lists the experimentally obtained values of the internal actions (bending moments and shear forces), describes the failure mode, and indicates the location of failure of the specimens. Table 7.10 shows the code predictions of shear capacity of characteristic portions of the specimens, whereas Table 7.11 contains the ratios of these values to their experimental counterparts. The figures show both the load–displacement curves and the mode of failure and associated crack pattern of the specimens. Although the work focuses primarily on

Table 7.8 Experimentally established and calculated values of the transverse force (in kN) and corresponding displacement (in mm) at various load levels for the specimens tested

Specimen	Experimental					Calculated						
	P_{max}	δ_{Pmax}	$\delta_{0.85max}$	δ_{sust}	δ_{fail}	P_y	$P_{ny}=P_{1P}$	P_{2P}	P_{max}/P_{2P}	δ_{ny}	μ_{sust}	μ_{fail}
ACI-D16-M	178	29.0	61.2	–	–	109	144	166	1.07	10.9	5.5	–
EC-D16-M	164	34.2	61.6	–	–	101	137	157	1.04	11.6	5.5	–
CFP-D16-M	154	29.7	55.8	–	–	97	133	152	1.01	9.5	5.9	–
ACI-D16-C	171	28.3	–	45.8	–	109	144	166	1.03	11.0	4.2	–
EC-D16-C	167	31.1	–	17	34	101	137	157	1.06	11.6	1.5	2.9
CFP-D16-C	162	30.7	–	41.4	51.5	97	133	152	1.08	9.5	4.4	5.4

P_y, P_{1P}, P_{2P} and P_{max}: the values of transverse force at first (beginning of) yield, 1st plastic hinge, 2nd plastic hinge (predicted load-carrying capacity) and experimentally established peak level, respectively; δ_{ny}, δ_{Pmax}, $\delta_{0.85Pmax}$, δ_{sust}, and δ_{fail}: the values of transverse displacement at P_y, P_{max}, the post-peak value of $P = 0.85P_{max}$, the maximum sustained loading cycle and loading cycle that caused failure, respectively; $\mu_{sust} = \delta_{0.85Pmax}/\delta_{ny}$ or $\mu_{sust} = \delta_{sust}/\delta_{ny}$ for the cases of monotonic and cyclic, respectively, loading, and $\mu_{fail} = \delta_{fail}/\delta_{ny}$ for the case of cyclic loading, $\delta_{1P} = \delta_{ny}$

Table 7.9 Experimental values of bending moment (in kNm) at support B ($M_{B, e}$) and point load at C ($M_{C, E}$) and comparison with their design values ($M_{B, f}$ and $M_{C, f}$), shear force (in kN) within portions AB ($V_{AB, E}$), BC ($V_{BC, E}$) and CD ($V_{CD, E}$), and mode of failure (and its location) for all specimens (values in **bold** indicate locations of shear failure), (locations of B, C, AB, BC and CD as in Fig. 7.24)

Experimental values									
Specimen	$M_{B, E}$	$M_{B, E}/M_{B, f}$	$M_{C, E}$	$M_{C, E}/M_{C, f}$	$V_{AB, E}$	$V_{BC, E}$	$V_{CD, E}$	Mode of failure	Location of failure
ACI-D16-M	63	1.17	64	1.18	53	121	57	Flexural	Load point C
EC-D16-M	54.8	1.02	60.7	1.19	46	110	54	Flexural	Load point C
CFP-D16-M	51.1	1.02	57.9	1.16	43	103	50	Flexural	Load point C
ACI-D16-C	57.2	1.06	63.3	1.17	48	114	56	Flexural	Load point C
EC-D16-C	55.8	1.09	61.0	1.20	47	**112**	55	Web horizontal cracking	Middle of BC
CFP-D16-C	54.2	1.08	61.1	1.22	45.0	109	53	Comp. zone failure and inclined cracking	CD right of C

structural behaviour under cyclic loading, testing under monotonic loading is considered essential for purposes of comparison. Moreover, the results obtained under monotonic loading are used to define a nominal value of the yield point, which forms the basis for the assessment of the ductility ratio of all specimens subsequently tested under cyclic loading.

Table 7.10 Shear capacities (in kN) predicted by EC2/EC8 and ACI for the various portions of the specimens tested (values in **bold** indicate locations of shear failure), (element portions AB, BC, CD as in Fig. 7.24.)

Specimen	Specimen portion											
	AB left side		AB right side		BC both ends		BC middle		CD left side		CD right side	
	EC	ACI	EC	ACI	EC	ACI	EC	ACI	EC	ACI	EC	ACI
ACI-D16-M	168	192	256	288	256	288	168	192	256	288	168	192
EC-D16-M	123	144	245	274	245	274	123	144	245	274	123	144
CFP-D16-M	108	130	100	130	100	130	152	174	100	130	108	130
ACI-D16-C	168	192	256	288	256	288	168	192	256	288	168	192
EC-D16-C	123	144	245	274	245	274	**123**	**144**	245	274	123	144
CFP-D16-C	108	130	100	130	100	130	152	174	100	130	108	130

Table 7.11 Ratios of shear capacities predicted by EC2/8 and ACI to measured shear forces for the various portions of the specimens tested (values in **bold** indicate locations of shear failure)

Specimen	Specimen portion											
	AB left side		AB right side		BC both ends		BC middle		CD left side		CD right side	
	EC2	ACI	EC2	ACI	EC2	ACI	EC2	ACI	EC2	ACI	EC2	ACI
ACI-D16-M	3.20	3.65	4.87	5.48	2.11	2.37	1.38	1.58	4.52	5.08	2.96	3.39
EC-D16-M	2.69	3.15	5.36	6.00	2.22	2.49	1.12	1.31	4.55	5.08	2.28	2.67
CFP-D16-M	2.53	3.05	2.35	3.05	0.97	1.27	1.48	1.69	1.99	2.59	2.15	2.59
ACI-D16-C	3.53	4.03	5.37	6.05	2.23	2.51	1.46	1.67	4.56	5.13	2.99	3.42
EC-D16-C	2.64	3.10	5.27	5.89	2.18	2.44	**1.10**	**1.28**	4.46	4.99	2.24	2.62
CFP-D16-C	2.39	2.88	2.22	2.88	0.92	1.20	1.40	1.60	1.88	2.44	2.03	2.44

Element portions AB, BC, CD as in Fig. 7.24

7.5.3 Discussion of Results

(a) Monotonic loading

Figure 7.27 shows the load–displacement curves obtained for the specimens tested under monotonic loading. On these figures, the location of the nominal yield point used for assessing the specimen ductility ratio is also indicated. The location of this point is determined as for the case of the beam-column elements discussed in Sect. 7.2.3.

It is evident from the figures that all specimens monotonically loaded exhibit ductile behaviour. In fact, Table 7.8 indicates that the average value of the ductility ratio (μ) of the specimens, defined as the ratio of the displacement at a post-peak load of 85 % the load-carrying capacity ($\delta_{0.85Pmax}$) to the displacement at nominal yield (δ_{ny}), i.e. $\mu = \delta_{0.85Pmax}/\delta_{ny}$, is over 5. Moreover, the table shows that, for all specimens the experimental values of load-carrying capacity either equal or exceed their calculated design counterparts.

Finally, it can be seen from the load–displacement characteristics (Fig. 7.27) that testing is stopped at a relatively large ductility, before significant loss of load-carrying capacity. In all cases, the crack patterns shown in Fig. 7.28 are characterised by flexural cracking.

Fig. 7.27 Load-displacement curves for D14 specimens tested under monotonic loading: **a** ACI-D14-M; **b** EC-D14-M; **c** CFP-D14-M [the *triangular symbols* represent (*moving upwards*) the points at first yield, first plastic hinge and second plastic hinge, respectively]

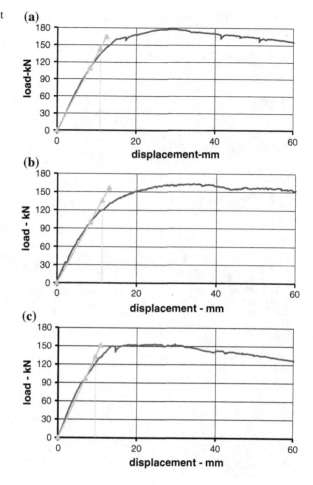

(b) Cyclic loading

Figure 7.29 shows that the behaviour of the specimens tested under cyclic loading is not as ductile as that of the specimens subjected to monotonic loading. In fact, Table 7.8 shows that, for the loading cycle that induces the maximum sustained displacement (δ_{sust}), the ductility ratio ($\mu_{sust} = \delta_{sust}/\delta_{ny}$) varies from 1.5 to 4.4, while the ductility ratio at failure ($\mu_{fail} = \delta_{fail}/\delta_{ny}$, where δ_{fail} is the displacement at which failure occurs) only once exceeds 5.

It is interesting to note in Table 7.8 that a significantly lower ductility is exhibited by all specimens designed in accordance with EC2/EC8. Figure 7.29b shows that such low ductility is characterised by a brittle type of failure due to near horizontal splitting of the portion of the specimen between the point load and the middle support (portion BC in Fig. 7.24), i.e. the region of the point of contraflexure. This can also be seen from Table 7.11, which shows that failure in the middle (BC) portion of the specimen (indicated by bold numbers) occurs despite the apparent margins of

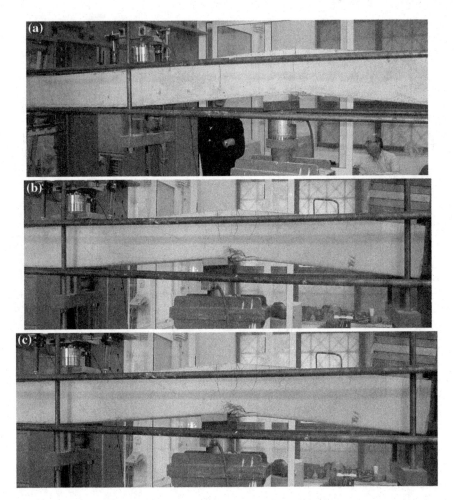

Fig. 7.28 Modes of failure and associated crack patterns for D14 specimens tested under monotonic loading: **a** ACI-D14-M; **b** EC-D14-M; **c** CFP-D14-M

safety against shear failure in both ACI (34 %) and EC2 (11 %). However, for specimen ACI-D16-C, an increase of the safety margin to the level specified by the CFP method appears to safeguard against horizontal splitting in the region of the point of inflection of portion BC. Naturally, such horizontal splitting does not occur in the specimens designed to the CFP method (see Fig. 7.30c).

7.5.4 Concluding Remarks

Under cyclic loading, the specimens designed to current code provisions suffer premature failure due to near-horizontal web cracking in the region of the point of contraflexure.

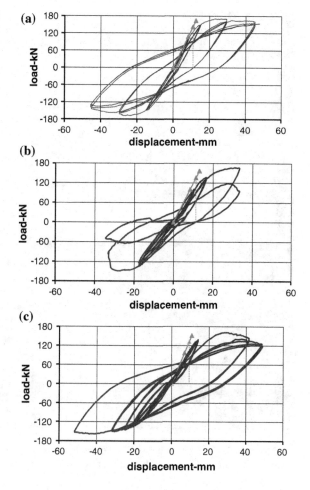

Fig. 7.29 Load-displacement curves for D16 specimens tested under cyclic loading: **a** ACI-D16-C; **b** EC-D16-C; **c** CFP-D16-C [the *triangular symbols* represent (*moving upwards*) the points at yield, first plastic hinge and second plastic hinge, respectively]

The above mode of failure is found to be prevented by increasing the amount of stirrups in the region of the point of contraclecture to the levels specified by the CFP method.

The specimens designed to the CFP method appear to be more likely to satisfy the performance requirements of current codes for strength and ductility.

7.6 Conclusions

- Unlike the methods adopted by current codes for the design of concrete structures, the proposed method is expressed in a unified form which is applicable to all structural elements under the actions considered in current design practice without the need of modification.

Fig. 7.30 Modes of failure and associated crack patterns for D14 specimens tested under cyclic loading: **a** Specimen ACI-D14-C; **b** specimen EC-D14-C; **c** specimen CFP-D14-C

- The proposed method is found to produce design solutions which satisfy the requirements of current codes for structural performance of earthquake-resistant structures in all cases investigated; these cases include structural members such as beams, columns, walls, beam-to-beam or column-to-column connections, and beam-to-column joints.
- Unlike the methods adopted by current codes, the proposed method classifies the regions of points of contraflexure (beam-to-beam or column-to-column connections) as critical and specifies reinforcement, when needed, for safeguarding against the brittle types of failure suffered in recent earthquakes by the vertical elements of RC structures designed by current codes.

References

1. Kotsovos GM, Kotsovos DM, Kotsovos MD, Kounadis A (2011) Seismic design of structural concrete walls: an attempt to reduce reinforcement congestion. Mag Concr Res 63(4):235–245
2. Kotsovos MD, Pavlovic MN (2001) The 7/9/99 Athens earthquake: causes of damage not predicted by structural-concrete design methods. Struct Eng 79(15):23–29
3. Kotsovos GM (2011) Seismic design of RC beam-column elements. Mag Concr Res 63(7):527–537
4. Kotsovou G, Mouzakis H (2012) Seismic design of RC external beam-column joints. Bull Earthq Eng 10(2):645–677
5. EN 1998-1 (2004) Eurocode 8: design of structures for earthquake resistance—part 1: general rules, seismic actions and rules for buildings
6. Kotsovos GM, Vougioukas E, Kotsovos MD (2003) Reducing steel congestion without violating the seismic performance requirements. ACI Struct J 100(1):11-18
7. Kotsovos MD, Baka A, Vougioukas E (2003) Earthquake-resistant design of reinforced-concrete structures: shortcomings of current methods. ACI Struct J 100(1):11–18
8. Jelic I, Pavlovic MN, Kotsovos MD (2004) Performance of structural-concrete members under sequential loading and exhibiting points of inflection. Comput Concr 1(1):99–113
9. EN 1992-1 (2004) Eurocode 2: design of concrete structures—part 1-1: general rules and rules for buildings
10. Zygouris NSt, Kotsovos GM, Kotsovos MD Effect of transverse reinforcement on short structural wall behavior. Mag Concr Res 65(17):1034-1043
11. Carpaer KL (1998) Current structural safety topics in North America. Struct Eng 76(12):233–239
12. Japanese Society of Civil Engineers (1995) Preliminary report on the Great Hanshin earthquake, 17 Jan 1995, p 138
13. Collins MP, Vecchio FJ, Selby RG, Gupta PR (1997) The failure of an offshore platform. Concr Int 19(8):28–34
14. Kotsovos MD, Pavlovic MN (1999) Ultimate limit-state design of concrete structures: a new approach. Thomas Telford, London, p 164
15. Kotsovos MD, Baka A, Vougioukas E (2003) Earthquake-resistant design of reinforced-concrete structures: shortcomings of current methods. ACI Struct J 100(1):11–18
16. Jelic I, Pavlovic MN, Kotsovos MD (2004) Performance of structural-concrete members under sequential loading and exhibiting points of inflection. Comput Concr 1(1):99–113
17. Kotsovos GM, Zeris C, Pavlovic MN (2005) Improving RC seismic design through the CFP method. Proc ICE Struct Buildings 158(SB5):291–302
18. Kotsovos GM, Zeris C, Pavlovic MN (2007) Earthquake-resistant design of indeterminate reinforced-concrete slender column elements. Eng Struct 29:163–175

Chapter 8
Design Applications

8.1 Introduction

The present chapter presents a number of design examples of the application of the compressive force-path (CFP) method discussed in the preceding chapters, with particular emphasis being given to design applications for earthquake-resistant structures. For purposes of comparison, the cases presented also include design solutions obtained from methods adopted by current codes. In all cases, all safety factors are set equal to 1; for application in practice, the code recommended values may be used.

8.2 Column Exhibiting Type II Behaviour

The column illustrated in Fig. 8.1 is part of a frame structure; it has a square cross section with 500 mm side, a clear height of 2,350 mm and longitudinal reinforcement comprising twelve 28 mm diameter bars arranged symmetrically about the axes of symmetry of the cross section at a spacing of 140 mm along the side faces of the column, with their geometric centre lying at a distance of 40 mm from the column faces. The uniaxial cylinder compressive strength of concrete is 30 MPa, whereas the yield stress of the steel is 550 MPa. In the following, the CFP method is used to determine the amount and arrangement of the transverse reinforcement required for safeguarding against brittle types of failure when the bending moment and the shear force acting at the column ends combine with an axial force $N = 1,500,000$ N. The transverse reinforcement specified by the European codes of practice EC2 and EC8 is also presented for purposes of comparison.

M. D. Kotsovos, *Compressive Force-Path Method*, Engineering Materials,
DOI: 10.1007/978-3-319-00488-4_8, © Springer International Publishing Switzerland 2014

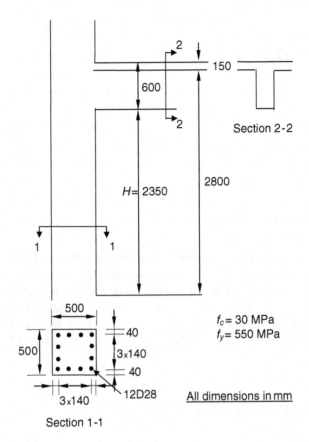

f_c = 30 MPa
f_y = 550 MPa

All dimensions in mm

Fig. 8.1 Geometric characteristics, longitudinal reinforcement details and material characteristics of slender column

8.2.1 CFP Design

Flexural capacity. Figure 8.2 depicts the internal actions that would develop at the end cross-section were the column capable to reach its flexural capacity. Expression 3.3, in which $f_t = 2.37$ MPa is obtained from expression 3.2a,

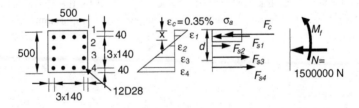

Fig. 8.2 Internal actions developing at the end cross-sections of the column in Fig. 8.1

yields $\sigma_a = 41.85$ MPa and thus the force sustained by concrete in the compressive zone is $F_c = \sigma_a bx = 41.85 \times 500x = 20{,}925x$, where b is the cross section width and x the depth of the compressive zone. If it is assumed that the longitudinal bars of layers 1 and 2 (see Fig. 8.2) remain within their elastic range of behaviour when flexural capacity is attained, then the forces sustained by the bars are $F_{s1} = A_{s1}f_{s1} = A_{s1}\varepsilon_{s1}E_s$ and $F_{s2} = A_{s2}f_{s2} = A_{s2}\varepsilon_{s2}E_s$, respectively ($E_s = 200{,}000$ MPa is the modulus of elasticity of the steel, whereas A_{s1}, f_{s1}, ε_{s1}, and A_{s2}, f_{s2}, ε_{s2} are the total cross-sectional areas of the steel bars in layers 1 and 2 and the corresponding stresses and strains). Since from the compatibility conditions $\varepsilon_{s1} = 0.0035 \times (x - 40)/x$ and $\varepsilon_{s2} = 0.0035 \times (x - 180)/x$, $F_{s1} = 4 \times (\pi \times 28^2/4) \times 200{,}000 \times 0.0035 \times (x - 40)/x = 1{,}724{,}106 \times (x - 40)/x$ and $F_{s2} = 2 \times (\pi \times 28^2/4) \times 200{,}000 \times 0.0035 \times (180 - x)/x = 862{,}053 \times (180 - x)/x$. On the other hand, assuming that the steel bars of layers 3 and 4 are at yield when flexural capacity is attained, the forces sustained by these bars are $F_{s3} = A_{s3}f_y$ and $F_{s4} = A_{s4}f_y$, where f_y is the yield stress of the steel and A_{s3}, A_{s4} are the total cross-sectional areas of the bars of layers 3 and 4, respectively. Thus, the total forces sustained by the steel bars are $F_{s3} = 2 \times (\pi \times 28^2/4) \times 550 = 677{,}327$ N and $F_{s4} = 4 \times (\pi \times 28^2/4) \times 550 = 1{,}354{,}655$ N.

Considering the equivalence between the longitudinal internal and external actions, $F_c + F_{s1} - F_{s2} - F_{s3} - F_{s4} = N$, in which F_c, F_{s1}, F_{s2}, F_{s3}, and F_{s4} are replaced with their expressions (functions of x only), yields $x \approx 129$ mm and thus, $F_c = 2{,}699{,}325$ N, $F_{s1} = 1{,}189{,}500$ N, $F_{s2} = 340{,}812$ N, $F_{s3} = 677{,}327$ N, and $F_{s4} = 1{,}354{,}655$ N, the direction of the forces being as indicated in Fig. 8.2. Replacing the values of x and of the internal longitudinal forces into equation $M_f = F_c(250 - 0.5x) + F_{s1}(250 - 40) - F_{s2}(250 - 180) + F_{s3}(250 - 180) + F_{s4}(250 - 40)$, which expresses the equivalence between the moments of the internal and external forces, yields $M_f \approx 1{,}058 \times 10^6$ Nmm. Then, the shear force corresponding to M_f is $V_f = 2 M_f/H = 2 \times 1{,}058 \times 10^6/2{,}350 \approx 900 \times 10^3$ N.

Physical model. Since the column is essentially equivalent to a beam subjected to similar end actions (longitudinal and transverse forces coupled with a moment), it is modelled as discussed in Sect. 6.2.1, with its model being illustrated in Fig. 8.3. For a column with the same cross-sectional characteristics throughout its height H, the point of contraflexure (point of zero bending moment) forms at mid height (location 3 in Fig. 8.3), and hence its two constituent 'cantilevers' have a shear span $a_v = H/2 = 2{,}350/2 = 1{,}175$ mm and a shear span-to-depth ratio $a_v/d = 1{,}175/380 \approx 3.1 > 2.5$, thus exhibiting type II behaviour ($d \approx 380$ mm being the distance of the point of application of the resultant of the tensile forces F_{s2}, F_{s3}, and F_{s4} sustained by the longitudinal reinforcement from the extreme compressive fibre). As discussed in Sect. 2.3, for type II behaviour, brittle failure may occur either in the region (indicated as location 1 in Fig. 8.3) where the horizontal and inclined elements of 'frame' join (at a distance of $2.5d = 2.5 \times 380 \approx 950$ mm from the 'internal' support forming at mid height) or in the region (indicated as location 2 in Fig. 8.3) where the bending moment and shear force attain their largest values. [The causes of these types of behaviour are fully discussed in Chap. 2 (Sect. 2.3)].

Fig. 8.3 Physical model
of the column in Fig. 8.1
indicating locations of
possible brittle failure

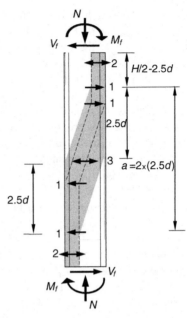

Predominantly ompressive stress field

Transverse tension at column ends (2) and mid
height (3)

Transverse tension at locations of change in
compressive force path direction (1)

However, as discussed in Sect. 6.2.1, brittle failure of the column may also
occur within its middle region (shaded grey in Fig. 8.3) where the compressive
force developing on account of bending combined with axial force is transferred
from the upper right-hand side end of the column to its lower left-hand side end.
As indicated in Fig. 8.3, the length a of the above region is equal to the distance
between locations 1 of the upper and lower 'cantilever' members of the column
(i.e. $a = 2 \times (2.5d) = 2 \times (2.5 \times 380) \approx 1{,}900$ mm). Moreover, since, as also discussed in Sect. 6.2.1, $a = 1{,}900$ mm $> 2.5d \approx 950$ mm, the slope of the inclined
compression is 1:2.5, with either of, or both, locations 1 moving towards the column mid-height so as both the slope of the inclined compression is maintained
constant and the internal force conditions (zero bending moment and concentric
axial force resultant) at the location of the point of contraflexure (indicated as
location 3 in Fig. 8.3) are not violated.

Transverse reinforcement at locations 1. From expression 3.11, the 'shear'
force that can be sustained without the need of transverse reinforcement in the
region of a location 1 is $V_{II,1} = 0.5f_tbd(h - x_o)/(h - x) = 0.5 \times 2.37 \times 500 \times 3$
$80 \times (500 - 91)/(500 - 129) = 248{,}911$ N $< 900{,}000$ N, where $x_o = 91$ mm is
the depth of the compressive zone for $N = 0$. (Note that x_o results by considering the equivalence between longitudinal internal and external forces indicated in

Fig. 8.2 ($F_c + F_{s1} - F_{s2} - F_{s3} - F_{s4} = 0$) in which F_c, F_{s1}, F_{s2}, F_{s3}, and F_{s4} are replaced with their expressions which are functions of x_o only). Thus, there is a need for transverse reinforcement ($A_{sv,1}$) capable of sustaining the whole transverse tensile force $T_{II,1} = V_{II,1} = V_f = 900,000$ N developing within a length $2d = 2 \times 380 \approx 760$ mm, i.e. $A_{sv,1} = 900,000/550 = 1,636.364$ mm^2. Such reinforcement is placed within a length extending to a distance $d = 380$ mm on either side of location 1. This amount of reinforcement is equivalent to four-legged stirrups of 8 mm diameter at a spacing of 93.38 mm; the latter being rounded up to 90 mm (four-legged D8 @ 90) for practical purposes. However, as the distance of location 1 from the support closest to it is $H/2 - 2.5d = 1,175 - 2.5 \times 380 \approx 225$ mm $< d \approx 380$ mm (see Fig. 8.4), the stirrup spacing within this length should be $93.38 \times 225/380 = 55.3$ mm, rounded up to 50 mm. Moreover, the reinforcement (four legged D8 @ 90) required in the region of location 1 is placed throughout the region where the compressive force is likely to change direction, since it is inevitable for locations 1 to move towards the column mid height for the reasons discussed earlier.

Transverse reinforcement at locations 2. From expression 3.13, the 'shear' force that can be sustained without the need of transverse reinforcement in the region of location 2 is $V_{II,2} = F_c*[1 - 1/(1 + 5|f_t|/f_c)] = 2,689,574 \times [1 - 1/(1 + 5|2.37|/30)] = 761,564$ N $< V_f = 900,000$ N. Thus, there is need for transverse reinforcement capable of sustaining the tensile stresses developing in the region of locations 2 due to the loss of bond between concrete and the longitudinal reinforcement. The amount of reinforcement required for this purpose is calculated through the use of expressions 4.2 and 4.3. From expression 4.2, the nominal value of the transverse tensile stresses developing within the region of locations 2

Fig. 8.4 Reinforcement details of column of Fig. 8.1 in accordance with CFP method

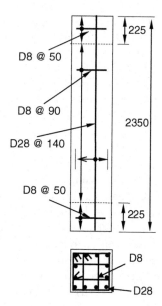

(extending from location 1 (i.e. the location of the joint of the vertical and inclined elements of the 'frame') to the column end) is $\sigma_t = 30/[5 \times (2{,}689{,}574/896{,}936 - 1)] \approx 3$ MPa. Then, from expression 4.3a, the largest of the tensile stress resultants developing within this region (with a length of 225 mm) in the directions of the cross-sectional axes of symmetry is $T_{II,2v} = 3 \times 500 \times 225/2 \approx 168{,}750$ N. The amount of transverse reinforcement required to sustain $T_{II,2v}$ is $A_{sv,2v} = 168{,}750/550 = 306.82$ mm^2, which is equivalent to one four-legged 8 mm diameter stirrup. However, the need for such reinforcement is covered by the reinforcement already specified within the same region for safeguarding against brittle failure at locations 1. On the other hand, the tensile stress resultant developing in the direction orthogonal to that of $T_{II,2v}$ is $T_{II,2h} = 3 \times 129 \times 225/2 = 43{,}537.5$ N. The amount of reinforcement required to sustain this force is $A_{sv,2h} = 43{,}537.5/550 = 79.16$ mm^2 which is also covered by the stirrup legs across the compressive zone specified for safeguarding against failure at locations 1.

Total transverse reinforcement. Therefore, in order to safeguard against brittle non-flexural types of failure, the transverse reinforcement used comprises of four-legged stirrups placed at spacing of 50 mm, within the regions extending to a distance of 225 mm from the column ends, and at spacing of 90 mm throughout the remainder middle region of the column.

8.2.2 EC2/EC8 Design

Figure 8.5 shows the amount and arrangement of transverse reinforcement specified by the European codes EC2 and EC8 for the column in Fig. 8.1 to exhibit either medium or high ductility. For both cases of ductility demand, the figure shows that the codes specify a significantly denser spacing within the end regions than within the middle region of the column, where the spacing provided is sufficient in accordance with the EC2 specifications (clause 6.2.3) for safeguarding against shear failure. The denser spacing within the end regions, termed 'critical lengths', is specified by EC8 [clauses 5.4.3.2.2 (medium ductility) and 5.5.3.2.2 (high ductility)] in order to safeguard the intended ductility demand. The figure shows that the code specifies both denser reinforcement spacing and longer critical lengths for the case of a high ductility demand.

Comparing the code specified transverse reinforcement details with those resulting from the application of the CFP method, it becomes apparent that, in accordance with the CFP method, the code specified amount of transverse reinforcement is inadequate for preventing the types of failure linked with type II behaviour both within the middle region of the column and the critical lengths, in the latter for the case of medium ductility demand only. On the other hand, the amount of transverse reinforcement specified by the code within the critical lengths for the case of a high ductility demand is larger than the amount required in accordance with the CFP method, since it spreads over a longer distance from the column ends (750 mm against 225 mm).

Fig. 8.5 Reinforcement
details of column of Fig. 8.1
in accordance with EC2/EC8
codes

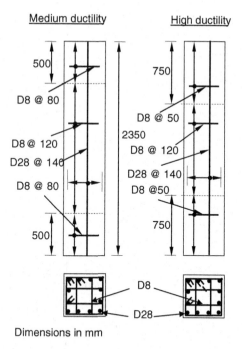

Dimensions in mm

8.3 Column Exhibiting Types of Behaviour II and III

The column illustrated in Fig. 8.6 differs from that in Fig. 8.1 in that its square
cross section has a 700 mm side, the longitudinal reinforcement spacing is
204 mm along the side faces of the column, with their geometric centre being at
a distance of 44 mm from the column faces, and the bending moment and shear
force acting at its ends are combined with an axial force $N = 2,900,000$ N. As for
the case of the column in Fig. 8.1, in the following, the CFP method is used to
determine the amount and arrangement of the transverse reinforcement required to
safeguard against brittle types of failure, with the transverse reinforcement specified by the European codes of practice EC2 and EC8 being also provided for comparison purposes.

8.3.1 CFP Design

Flexural capacity. Figure 8.7 depicts the internal actions that would develop at the
end cross-sections, were the column capable to reach its flexural capacity. By following the reasoning adopted for the calculation of flexural capacity in Sect. 8.2.1,
these internal actions are expressed as follows: $F_c = \sigma_a bx = 41.85 \times 700x =
292,950x$, $F_{s1} = 4 \times (\pi \times 28^2/4) \times 200,000 \times 0.0035 \times (x - 40)/x = 1,724,106 \times$

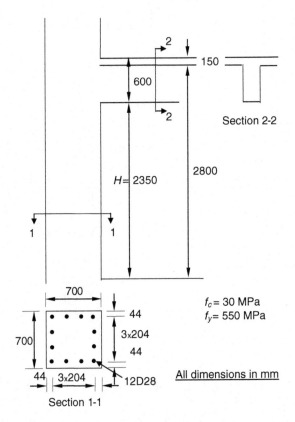

Fig. 8.6 Geometric characteristics, longitudinal reinforcement details and material characteristics of short column

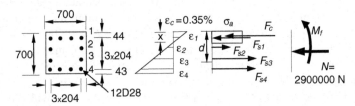

Fig. 8.7 Internal actions developing at the end cross-sections of the column in Fig. 8.6

$(x - 44)/x$, $F_{s2} = 2 \times (\pi \times 28^2/4) \times 200,000 \times 0.0035 \times (248 - x)/x = 862,050 \times (248 - x)/x$, $F_{s3} = 2 \times (\pi \times 28^2/4) \times 550 = 677,327$ N and $F_{s4} = 4 \times (\pi \times 28^2/4) \times 550 = 1,354,655$ N.

When replacing F_c, F_{s1}, F_{s2}, F_{s3}, and F_{s4} with the above expressions, equation $F_c + F_{s1} - F_{s2} - F_{s3} - F_{s4} = N$, which describes the equivalence between the longitudinal internal and external actions, yields $x \approx 147$ mm and thus, $F_c = 4,306,365$ N, $F_{s1} = 1,208,047$ N, $F_{s2} = 592,295$ N, $F_{s3} = 677,325$ N, and $F_{s4} = 1,354,650$ N, the force direction being as indicated in Fig. 8.7. Replacing the values of x, F_c, F_{s1}, F_{s2}, F_{s3} and F_{s4} in expression $M_f = F_c \times (350 - 0.5x) + F_{s1} \times (350 - 44) - F_{s2} \times (350 - 248) + F_{s3} \times (350 - 248) + F_{s4} \times (350 - 44)$, which describes the equivalence between the moments of the internal and external actions, yields $M_f \approx 1{,}985 \times 10^6$ Nmm. Then, the shear force corresponding to M_f is $V_f = 2\,M_f/H = 2 \times 1{,}985 \times 10^6/2{,}350 \approx 1{,}690 \times 10^3$ N.

Physical model. As for the column in Fig. 8.1, the column in Fig. 8.6 is modelled as discussed in Sect. 6.2.1, with its model being illustrated in Fig. 8.8. For a column with the same cross-sectional characteristics throughout its height H, the point of contraflexure forms at mid height (location 3 in Fig. 8.8); thus, the column's two constituent 'cantilevers' exhibit type III behaviour, since their shear span-to-depth ratio $a_v/d = 1{,}175/511.49 \approx 2.3 < 2.5$, where $a_v = H/2 = 2{,}350/2 = 1{,}175$ mm and $d \approx 511$ mm is the distance of the point of application of the resultant of the tensile forces F_{s2}, F_{s3}, and F_{s4} sustained by the longitudinal reinforcement from the extreme compressive fibre. As discussed in Sect. 2.3, for type III behaviour, non-flexural failure of the beam results from the reduction of the depth of the compressive zone owing to the extension of the inclined crack which is deeper than the flexural cracks. (The causes of this type of failure—occurring as a result of failure of the horizontal element of the 'frame', in the region of the joint of the horizontal and inclined elements of the 'frame'—are described in Sect. 2.3 of Chap. 2.)

Transverse reinforcement at locations 1′. The reduction in depth causes a reduction in flexural capacity and the maximum bending moment M_{III} that can be

Fig. 8.8 Physical model of the column in Fig. 8.6 indicating locations of possible brittle failure

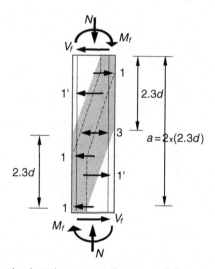

Predominantly compressive stress field
→ Transverse tension at mid height (3)
← Transverse tension at locations of change in
→ compressive force path direction (1) or where resultant of transverse stirrup forces should act in order to prevent type III failure (1′)

sustained by the column is assessed, as described in Sect. 3.3.3, by linear interpolation between two values of M_{III} corresponding to $a_v/d = 1$ ($M_{III} = M_f$) and $a_v/d = 2.5$ ($M_{III} = M_{II}^{(2.5d)}$). Thus for $a_v/d = 2.5$, expression 3.9 yields $V_{II,1} = 0.5 \times 2.37 \times 700 \times 511 \times (700 - 90)/(700 - 147) = 467{,}565$ N [where $x_o \approx 90$ mm is the depth of the compressive zone when $N = 0$ calculated as for the case of the column in Fig. 8.1 (see Sect. 8.2.1)], whereas from expression 3.13 $V_{II,2} = 4{,}30$ $6{,}365[1 - (1 + 5 \times 2.37/30)] = 1{,}219{,}365$ N. Thus, as discussed in Sect. 3.3.3, $V_{II} = \min(V_{II,1}, V_{II,2}) = 467{,}565$ N and $M_{III}^{(2.5d)} = 2.5dV_{II} = 2.5 \times 511 \times 467{,}56$ $5 = 597 \times 10^6$ Nmm. As a result, for $a_v/d = 2.3$, expression 3.14 yields $M_{III} = 5$ $97 \times 10^6 + (1{,}985.413 \times 10^6 - 597 \times 10^6)(2.5 \times 511 - 1{,}175)/(1.5 \times 511) = 7$ 83×10^6 Nmm $< M_f \approx 1{,}985 \times 10^6$ Nmm; thus failure will occur before flexural capacity is exhausted.

As discussed in Sect. 4.3.2, this type of failure can be prevented by uniformly distributing transverse reinforcement in the form of stirrups within the whole length of the shear spans. The amount of the reinforcement required is assessed from expression 4.6 which yields $A_{sv} = 2(1{,}985 \times 10^6 - 783 \times 10^6)/$ $(1{,}175 \times 550) = 3{,}710.64$ mm^2 within a length of 1,175 mm; this reinforcement is equivalent to four-legged 8 mm diameter stirrups at a spacing of 63.7 mm, the latter being rounded up to 60 mm (four-legged D8 @ 60).

Transverse reinforcement at locations 1. However, as for the case of the column in Fig. 8.1, brittle failure may also occur within the region (shaded grey in Fig. 8.8) where the compressive force developing on account of bending combined with axial force is transferred from the upper right-hand side end of the column to its lower left-hand side end. Since the length of this transfer region $a = H = 2{,}350$ mm is larger than $2.5d = 2.5 \times 511 \approx 1{,}277.5$ mm, the slope of the inclined compression is 1:2.5, with either of, or both, locations of change in the path of the compressive force (locations 1 in Fig. 8.8) moving towards the location of the mid-column height so as both the slope of the inclined compression is maintained constant and the internal force conditions at the location of the point of contraflexure are not violated.

This is a type II failure which, as discussed Sect. 2.3, occurs when the tensile force, $T_{II,1} = V_{II,1}$, developing in the region of a location 1, cannot be sustained by concrete alone; from expression 3.11, $V_{II,1} = 0.5 f_t b d (h - x_o)/(h - x) = 0.5 \times 2.37 \times 700 \times 511 \times (700 - 90)/(700 - 147) = 467{,}565$ N $< V_f = 1{,}690 \times 10^3$ N, where $x_o \approx 90$ mm is the depth of the compressive zone when $N = 0$. Thus, there is a need for transverse reinforcement ($A_{sv,1}$) capable of sustaining the whole transverse tensile force $T_{11,1} = V_f = 1{,}690 \times 10^3$ N developing within a length $2d = 2 \times 511 = 1{,}022$ mm, i.e. $A_{sv,1} = 1{,}690 \times 10^3/550 = 3{,}072.73$ mm^2 within a length $2d = 1{,}022$ mm. This amount of reinforcement is equivalent to four-legged stirrups of 8 mm diameter at a spacing of 66.89 mm, the latter being slightly larger than the amount already found to be required to safeguard against type III failure.

Total transverse reinforcement—Therefore, the transverse reinforcement required to safeguard against brittle non-flexural types of reinforcement is four-legged stirrups at a spacing of 60 mm throughout the column height, as indicated in Fig. 8.9.

Fig. 8.9 Reinforcement
details of column of Fig. 8.6
in accordance with CFP
method

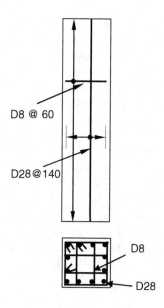

D8 @ 60

D28@140

D8

D28

8.3.2   EC2/EC8 Design

Figure 8.10 shows the amount and arrangement of transverse reinforcement specified by the European codes EC2 and EC8 for the column in Fig. 8.6 to exhibit
either medium or high ductility. The reinforcement arrangement is as for the
case of the column in Fig. 8.1; for both cases of ductility demands, EC8 (see
clauses 5.4.3.2.2 for medium ductility and 5.5.3.2.2 for high ductility) specifies a

Fig. 8.10 Reinforcement
details of column of Fig. 8.6
in accordance with EC2/EC8
codes

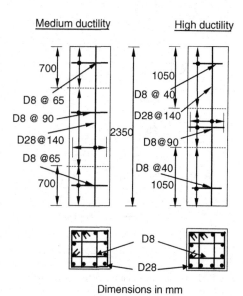

Medium ductility   High ductility

700

D8 @ 65

D8 @ 90

D28@140

D8 @65

700

2350

1050

D8 @ 40

D28@140

D8@90

D8 @40

1050

D8

D28

Dimensions in mm

significantly denser spacing within the end regions (critical lengths) than within the middle region of the column, where the spacing provided is sufficient in accordance with the EC2 specifications (clause 6.2.3) to safeguard against shear failure. As for the case of the column in Fig. 8.1, the figure shows that the code specifies both denser reinforcement spacing and longer critical lengths for the case of a high ductility demand.

Comparing the code specified transverse reinforcement details with those resulting from the application of the CFP method confirms the conclusions drawn for the case of the column discussed in Sect. 8.2: in accordance with the CFP method, the code specified amount of transverse reinforcement within the middle region of the column is inadequate for preventing the type II failure linked with the change in the path of the compressive force developing on account of bending; on the other hand, the amount of transverse reinforcement specified by the code within the critical lengths is slightly smaller, for the case of medium ductility, and significantly denser, for the case of high ductility demand, than the amount really required to satisfy the intended structural behaviour.

8.4 Coupling Beam of Type II Behaviour

The beam in Fig. 8.11 is considered to span the distance between two structural walls to which it is monolithically connected. The beam is intended to dissipate part of the energy in an easily repairable part of a structure comprising either a structural wall system or a dual structural wall-frame system subjected

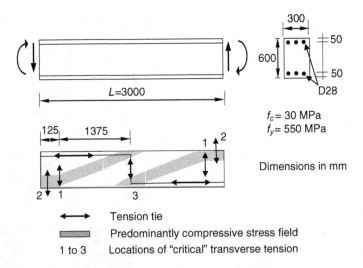

Fig. 8.11 Geometric characteristics, longitudinal reinforcement details and material characteristics (*top*) and physical model of coupling beam (*bottom*) exhibiting type II behaviour

to earthquake motion. As indicated in the figure, the beam has a clear span of 3,000 mm, a 600 mm high × 300 mm wide rectangular cross section, with a depth of 550 mm and longitudinal reinforcement comprising three 28 mm diameter top and bottom bars. The uniaxial cylinder compressive strength of concrete is 30 MPa, whereas the yield stress of the steel is 550 MPa. In the following, the proposed method is used to determine the amount and arrangement of the transverse reinforcement required for safeguarding against brittle types of failure. The transverse reinforcement specified by the European codes of practice EC2 and EC8 is also provided for purposes of comparison.

8.4.1 CFP Design

Flexural capacity. Figure 8.12 depicts the internal actions that would develop at the end cross-sections were the beam capable to reach its flexural capacity. From expression 3.3, in which $f_t = 2.37$ MPa is obtained from expression 3.2a, $\sigma_a = 41.85$ MPa and thus the force sustained by concrete in the compressive zone is $F_c = \sigma_a b x = 41.85 \times 300x = 12,555x$. If it is assumed that the longitudinal bars in compression remain within their elastic range of behaviour when flexural capacity is attained, then the total force sustained by these bars is $F_s' = A_s' f_s' = A_s' \varepsilon_s' E_s'$ where $E_s = 200,000$ MPa is the steel modulus of elasticity and As', f_s', ε_s' are the total cross-sectional area of, and the stresses and strains developing in, the steel bars in compression, respectively. Making use of the compatibility condition $\varepsilon_s' = 0.0035 \times (x - 50)/x$, the force sustained by the reinforcement in compression $F_s' = 3 \times (\pi \times 28^2/4) \times 200,000 \times 0.0035 \times (x - 50)/x = 1,293,080 \times (x - 50)/x$. On the other hand, assuming that the steel bars in tension are at yield when flexural capacity is attained, $F_s = A_s f_y$, where f_y is the steel yield stress and A_s is the total cross-sectional areas of the bottom. Thus, the total force sustained by these bars is $F_{s3} = 3 \times (\pi \times 28^2/4) \times 550 = 1,015,991$ N.

Replacing F_c, F_s', F_s with their expressions into expression $F_c + F_s' - F_s = 0$, which describes the equivalence between the longitudinal internal and external actions, yields $x \approx 61$ mm and thus, $F_c = 765,855$ N, $F_s' = 233,178$ N, $F_{s2} = 1,015,991$ N, the direction of the being as indicated in Fig. 8.12. Replacing x and the internal longitudinal forces with their values in equation $M_f = F_c(550 - 0.5x) + F_s (550 - 50)$, which expresses the equivalence between

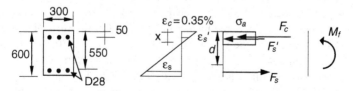

Fig. 8.12 Internal actions developing at the end cross-sections of the beam in Fig. 8.11

the moments of the internal and external actions, yields $M_f = F_c(550 - 0.5x) + F_s$ $(550 - 50)$ yields $M_f \approx 514 \times 10^6$ Nmm. Then, the shear force corresponding to M_f is $V_f = 2 M_f/L = 2 \times 514 \times 10^6/3,000 \approx 343 \times 10^3$ N.

Physical model. The beam is modelled as discussed in Sect. 6.2.1, with the model being illustrated in Fig. 8.11 (bottom). For a beam, such as that considered in the present case, which has the same cross-sectional characteristics throughout its length (L), the point of contraflexure forms at mid span (location 3 in Fig. 811 bottom), and thus its two constituent 'cantilevers' have a shear span-to-depth ratio $a_v/d = 1,500/550 \approx 2.73 > 2.5$ (where $a_v = L/2 = 3,000/2 = 1,500$ mm), thus exhibiting type II behaviour. As discussed in Sect. 2.3, for type II behaviour, brittle failure may either occur in the region of location 1 where the horizontal and inclined elements of 'frame' join (at a distance equal to $2.5d = 2.5 \times 550 = 1,375$ mm from mid span) or in the region of location 2 where bending moment and shear force attain their largest values. [The causes of these types of behaviour are fully discussed in Chap. 2 (Sect. 2.3)].

Transverse reinforcement at locations 1. From expression 3.9, the 'shear' force that can be sustained without the need of transverse reinforcement in the region of location 1 is $V_{II,1} = 0.5f_tbd = 0.5 \times 2.37 \times 300 \times 550 = 195,525$ N $<$ 343,000 N. Thus, there is a need for transverse reinforcement ($A_{sv,1}$) capable of sustaining the whole transverse tensile force $T_{II,1} = V_f = 343,000$ N developing within a length $2d = 2 \times 550 = 1,100$ mm, i.e. $A_{sv,1} = 343,000/550 = 623.64$ mm². This amount of reinforcement is required within a length extending to a distance $d = 550$ on either side of location 1, the latter being at a distance equal to $L/2 - 2.5 \ d = 3,000/2 - 1,375 = 125$ mm form the support closest to it. The reinforcement placed comprises five three-legged 8 mm diameter stirrups (with a total cross-sectional area $753.98 > 623.64$ mm²) at spacing of 220 mm, which is smaller than the maximum allowed spacing of $d/2 = 550/2 = 225$ mm (see item (f) of Sect. 4.4). It should also be noted that at least two of the stirrups are placed between the cross section at the support and location 1.

Transverse reinforcement at locations 2. From expression (3.13), the 'shear' force that can be sustained without the need of transverse reinforcement in the region of location 2 is $V_{II,2} = F_c^*[1 - 1/(1 + 5|f_t|/f_c)] = 765,855 \times [1 - 1/(1 + 5|2.37|/30)] = 216,855$ N $< V_f = 343,000$ N. Thus, there is need for transverse reinforcement capable of sustaining the tensile stresses developing in the region of location 2 due to the loss of bond between concrete and the longitudinal reinforcement. The amount of reinforcement required for this purposed is calculated through the use of expressions 4.2 and 4.3. From expression 4.2, the nominal value of the transverse tensile stresses developing within the region of location 2 (extending from location 1 (i.e. the location of the joint of the vertical and inclined elements of the 'frame') to the column end) is $\sigma_t = 30/[5 \times (765,855/343,000 - 1)] \approx 4.87$ MPa, whereas, from expression 4.3a, the stress resultant developing in the vertical direction within this region (with a length of 125 mm) is $T_{II,2v} = 4.9$ $4 \times 300 \times 125/2 \approx 91,313$ N. The amount of transverse reinforcement required to sustain $T_{II,2v}$ is $A_{sv,2v} = 91,313/550 = 166.02$ mm², which is less than the two three-legged 8 mm diameter stirrup (301.59 mm^2) already provided by the reinforcement required to safeguard against brittle failure at locations 1. On the other hand, the tensile stress resultant developing in the horizontal direction is $T_{II,2h} = 3 \times 61 \times 125/2 = 11{,}437.5$ N. The amount of reinforcement required to sustain this force is $A_{sv,2h} = 11{,}437.5/550 = 20.79$ mm^2 which is also covered by the stirrup legs across the compressive zone specified for safeguarding against failure at locations 1.

Transverse reinforcement at location 3. However, as discussed in Sect. 6.2.1, in the absence of an axial force brittle failure may also occur in the region of the point of contra-flexure [location 3 in Fig. 6.11 (bottom)] when the tensile strength of concrete is exceeded under the transverse tension developing in this region (see Fig. 6.2). From expression (3.8), the tensile force that can be sustained without the need of transverse reinforcement in the region of location 3 is $T_3 = 0.5 f_t bd = 0.5 \times 2.37 \times 300 \times 550 = 195{,}525$ N $< T_f = V_f = 343{,}000$ N. Thus, there is a need for transverse reinforcement ($A_{sv,3}$) capable of sustaining the whole transverse tensile force $T_f = 343{,}000$ N developing within a length $2d = 2 \times 550 = 1{,}100$ mm, i.e. $A_{sv,3} = 343{,}000/550 = 623.64$ mm^2. Such reinforcement is placed within a length extending to a distance $d = 550$ on either side of location 3; as for the case of the reinforcement in the region of location 1, the reinforcement provided comprises five three-legged 8 mm diameter stirrups at a spacing of 220 mm (see item (f) in Sect. 4.4).

Total transverse reinforcement. Therefore, the transverse reinforcement required to safeguard against brittle non-flexural types of failure comprises three-legged stirrups which, within the regions extending to a distance of 125 mm from the column ends, are placed at spacing of 60 mm, and throughout the remainder middle region of the beam, at spacing of 220 mm (see Fig. 8.13).

8.4.2 EC2/EC8 Design

Figure 8.14 shows the amount and arrangement of transverse reinforcement specified by the European codes EC2 and EC8 for the beam in Fig. 8.11 to exhibit either medium or high ductility. For both cases of ductility demands,

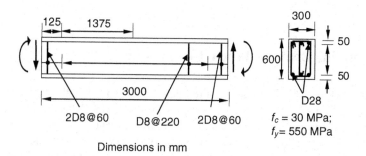

Dimensions in mm

Fig. 8.13 Reinforcement details of beam in Fig. 8.11 in accordance with CFP method

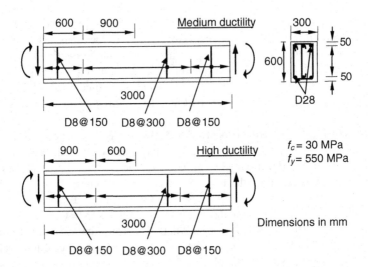

Fig. 8.14 Reinforcement details of beam in Fig. 8.11 in accordance with EC2/EC8

EC8 (see clauses 5.4.3.1.2 for medium ductility and 5.5.3.1.3 for high ductility) specifies a significantly denser spacing within the end regions (critical lengths) than within the middle region of the beam, where the spacing provided is sufficient in accordance with the EC2 specifications (clause 6.2.3) to safeguard against shear failure. Unlike the columns discussed in Sects. 8.2 and 8.3, the figure shows that while the code specifies longer critical lengths for the case of a high ductility demand, the stirrup spacing remains the same for both types of ductility demand.

Comparing the code specified transverse reinforcement details with those resulting from the application of the CFP method, it becomes apparent that, in accordance with the CFP method, the code specified amount of transverse reinforcement is inadequate for preventing the types of failure linked with type II behaviour within both the middle region and the critical lengths of the beam for both cases of ductility demand.

8.5 Coupling Beam of Type III Behaviour

The beam in Fig. 8.15 is similar in all respects but the span length, the latter being equal to 2,420 mm, with the beam in Fig. 8.11. As for the latter beam, the CFP method is used to determine the amount and arrangement of the transverse reinforcement required to safeguard against brittle types of failure, with the transverse reinforcement specified by the European codes of practice EC2 and EC8 being also provided for purposes of comparison.

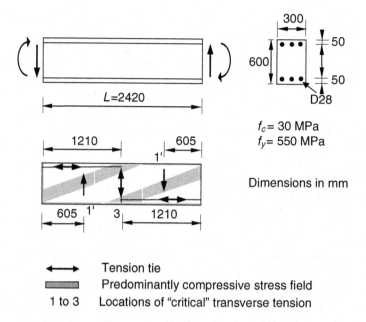

f_c = 30 MPa
f_y = 550 MPa

Dimensions in mm

⟷   Tension tie

▬   Predominantly compressive stress field

1 to 3   Locations of "critical" transverse tension

Fig. 8.15 Geometric characteristics, longitudinal reinforcement details and material characteristics (*top*) and physical model of coupling beam (*bottom*) exhibiting type III behaviour

8.5.1 CFP Design

Flexural capacity. Since the beam discussed in the present section has the cross-sectional characteristics of the beam discussed in Sect. 8.4, $M_f = 514 \times 10^6$ Nmm, with the shear force corresponding to M_f being $V_f = 2 M_f/L = 2 \times 514 \times 10^6/2,42$ $0 \approx 425 \times 10^3$ N.

Physical model. As for all structural elements discussed so far in the present chapter, the beam is modelled as discussed in Sect. 6.2.1, with its model being illustrated in Fig. 8.15 (bottom). Since the beam has the same cross-sectional characteristics throughout its length (L), the point of contra-flexure forms at mid span (location 3), and thus the two constituent 'cantilevers' have a shear span-to-depth ratio $a_v/d = 1,210/550 = 2.2 < 2.5$ (where $a_v = L/2 = 2,420/2 = 1,210$ mm), thus exhibiting type III behaviour. As discussed in Sect. 2.3, for type III behaviour, non-flexural failure of the beam results from the reduction of the depth of the compressive zone owing to the extension of the inclined crack which is deeper than the flexural cracks. (The causes of this type of failure—occurring as a result of failure of the horizontal element (reduced, in the present case, to a cross section at the beam ends) of the 'frame', in the region of the joint of the horizontal and inclined elements of the 'frame'—were described in Sect. 2.3 of Chap. 2.)

Transverse reinforcement at locations 1′. The reduction in depth causes a reduction in flexural capacity with the maximum bending moment M_{III} that can be sustained by the beam being assessed, as described in Sect. 3.3.3, by linear interpolation between two values of values of M_{III} corresponding to $a_v/d = 1$ ($M_{III} = M_f$) and $a_v/d = 2.5$ ($M_{III} = M_{II}^{(2.5d)}$). Thus for $a_v/d = 2.5$, expression 3.9 yields $V_{II,1} = 0.5 \times 2.37 \times 3$ $00 \times 550 = 195,525$ N, whereas from expression 3.13, $V_{II,2} = 765,855 \times [1 - 1/$ $(1 + 5 \times 2.37/30)] = 216,855$ N. Thus, as discussed in Sect. 3.3.3, $V_{II} = \min(V_{II,1},$ $V_{II,2}) = 195,525$ N and $M_{III}^{(2.5d)} = 2.5dV_{II} = 2.5 \times 550 \times 195,525 \approx 269 \times 10^6$ Nm m. As a result, for $a_v/d = 2.2$, expression 3.14 yields $M_{III} = 269 \times 10^6 + (514 \times 10^6$ $- 269 \times 10^6)(2.5 \times 550 - 1,210)/(1.5 \times 550) \approx 318 \times 10^6$ Nmm $< M_f \approx 514 \times 10^6$ Nmm; thus failure will occur before flexural capacity is exhausted.

As discussed in Sect. 4.3.2, this type of failure can be prevented by uniformly distributing transverse reinforcement in the form of stirrups within the whole length of the shear spans. The amount of such reinforcement required is assessed from expression 4.6 which yields $A_{sv} = 2(514 \times 10^6 - 318 \times 10^6)/$ $(1,210 \times 550) = 589.03$ mm² within a length of 1,210 mm which is equivalent to six two-legged 8 mm diameter stirrups at a spacing of 201.67 mm, the latter being rounded up to 200 mm (two-legged D8 @ 200).

Transverse reinforcement at location 3. However, as for the case of the beam discussed in Sect. 8.4, in the absence of an axial force brittle failure may also occur in the region of the point of contra-flexure [location 3 in Fig. 8.15 (bottom)] when the tensile strength of concrete is exceeded under the transverse tension developing in this region. From expression 3.8, the tensile force that can be sustained without the need of transverse reinforcement in the region of location 3 is $T_3 = 0.5f_tbd = 0.5 \times 2.37 \times 300 \times 550 = 195,525$ N $< T_f = V_f = 425,000$ N. Thus, there is a need for transverse reinforcement ($A_{sv,3}$) capable of sustaining the whole transverse tensile force $T_f = 425,000$ N developing within a length $2d = 2 \times 550 = 1,100$ mm, i.e. $A_{sv,3} = 425,000/550 = 772.72$ mm². Such reinforcement is placed within a length extending to a distance $d = 550$ on either side of location 3. This amount of reinforcement is equivalent to eight two-legged stirrups of 8 mm diameter at a spacing of 137.5 mm rounded up to 130 mm; the latter being smaller than the spacing of 200 already specified throughout the beam length in order to safeguard against type III failure.

Total transverse reinforcement. Therefore, the transverse reinforcement required to safeguard against brittle non-flexural types of failure comprises two-legged stirrups of 8 mm diameter at spacing of 130 mm within the middle portion of the beam extending 550 mm on either side of the mid cross section, and at spacing of 200 mm in the remainder of the beam as indicated in Fig. 8.16.

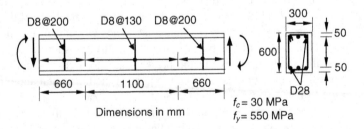

Fig. 8.16 Reinforcement details of beam in Fig. 8.15 in accordance with CFP method

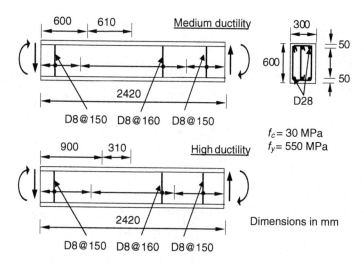

Fig. 8.17 Reinforcement details of beam in Fig. 8.15 in accordance with EC2/EC8

8.5.2 EC2/EC8 Design

Figure 8.17 shows the amount and arrangement of transverse reinforcement specified by the European codes EC2 and EC8 for the beam in Fig. 8.15 to exhibit either medium or high ductility. As for the case of the beam discussed in Sect. 8.4, for both cases of ductility demands, EC8 (see clauses 5.4.3.1.2 for medium ductility and 5.5.3.1.3 for high ductility) specifies a significantly denser spacing within the end regions (critical lengths) than within the middle region of the beam, where the spacing provided is sufficient in accordance with the EC2 specifications (clause 6.2.3) to safeguard against shear failure. The figure also shows that, for the case of a high ductility demand, the code specifies longer critical lengths, whereas the stirrup spacing is similar with that specified for the case medium ductility.

Comparing the code specified transverse reinforcement details with those resulting from the application of the CFP method shows that, in accordance with the CFP method, the code specified amount of transverse reinforcement is inadequate for preventing failure within the middle region of the beam, whereas, within the critical lengths, the code specified amount is larger than that deemed necessary by the CFP method for safeguarding against a non-flexural type of failure.

8.6 External Beam-Column Joint

For the beam-column joint in Fig. 8.18, the design details and material properties of the beam and column elements are those shown in Figs. 8.13 and 8.4, respectively. The figure also shows that the beam longitudinal bars are anchored at

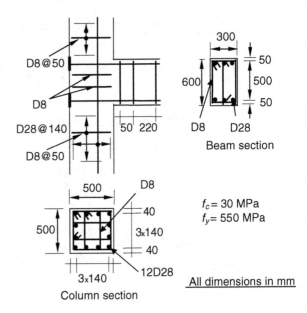

Fig. 8.18 External beam-column joint: Geometric characteristics, reinforcement details and material characteristics of joint, beam and column elements

Column section

All dimensions in mm

the back left-hand side end face of the joint through the use of steel end plates. In the following, the CFP method is used to determine the transverse reinforcement required to safeguard against the occurrence of significant cracking in the joint before the formation of a 'plastic' hinge in the adjacent end region of the beam. As in all design cases discussed in the preceding sections of the present chapter, the code specified reinforcement is also provided for purposes of comparison.

Referring to Fig. 6.15, the inclination a of the diagonal strut forming within the joint can be calculated from $sina = z_c/(z_c^2 + z_b^2)^{0.5}$, where z_c and z_b represent the distances between the points of application of the compressive force developing in concrete on account of bending of the column and beam elements, respectively, and the geometric centre of the tensile steel i.e. $z_c = d_c − x_c/2 = 380 − 1$ $29/2 = 315$ mm and $z_b = d_b − x_b/2 = 550 − 61/2 ≈ 519$ mm (with the subscripts 'b' and 'c' indicating beam and column, respectively). Thus, $sin\ a = 315/$ $(315^2 + 519^2)^{0.5} = 0.519$ (and, hence, $cosa = (1 − sin^2a)^{1/2} = 0.855$). As it is also indicated in Fig. 6.15, the angle b between the joint diagonal and the tangents at the ends of any of the two symmetrical (with respect to the joint diagonal) trajectories of the two compressive stress resultants, $F_j/2$, carried by the diagonal strut, can be determined from $tanb = (z_c/8)/[(z_c^2 + z_b^2)^{0.5}/2] = (z_c/4)/$ $(z_c^2 + z_b^2)^{0.5} = (315/4)/(315^2 + 519^2)^{0.5} ≈ 0.13$.

As discussed in Sect. 6.4.4, the horizontal component of the compressive force developing within the diagonal strut is equal to the resultant of the horizontal forces acting at the upper face of the joint, i.e. $F_{j,h} = F_{s,b} − V_c$, where $F_{s,b}$ is the tensile force developing in the longitudinal bars of the beam on account of bending, with $F_{s,b} ≈ 1{,}016 × 10^3$ N (see Sect. 8.4.1), and V_c is the shear force developing at the joint-upper column intersection when plastic hinge formation occurs in the beam region adjacent to the joint. Since from the moment equilibrium of the joint $M_{f,b} = 2M_c$ (with subscripts 'b' and 'c', as discussed earlier, indicating beam and column, respectively), an approximate estimate of V_c may be obtained by assuming that the end moments in the column are equal, even before the formation of plastic hinges at the column ends; then, $V_c = 2M_c/H = M_{f,b}/H = 514 \times 10^6/2,350 \approx 219 \times 10^3$ N. Therefore, $F_{j,h} = 1,016 \times 10^3 - 219 \times 10^3 = 797 \times 10^3$ N and thus, from expression 6.2, the compressive force carried by the diagonal strut is $F_j = F_{j,h}/\sin a = 797 \times 10^3/0.519 = 1,536 \times 10^3$ N, the latter value being smaller than the maximum force $F_{Rj,max}$ (obtained from expression 6.4) that can be sustained by the diagonal strut ($F_{Rj,max} = (315/3) \times 500 \times 30 = 1,575 \times 10^3$ N).

From expression 6.5, the tensile stress resultant developing across the diagonal strut $T_j = F_j \tan b = 1,536 \times 10^3 \times 0.13 \approx 199 \times 10^3$ N. If the latter force is decomposed into two components, one in the horizontal direction and one along the diagonal strut, then, the horizontal component $T_{j,h} = T_j/\cos a = 199 \times 10^3/0.855 \approx 234 \times 10^3$ N and, therefore, the amount of reinforcement, in the form of stirrups, required to sustain this stress resultant is $A_{sj,h} = T_j/f_y = 234 \times 10^3/550 = 425.31$ mm^2 (see expression 6.7), which is nearly equivalent to two four-legged stirrups with a total cross-sectional area of 402.13 mm^2, only slightly smaller than the required amount (see Fig. 8.18).

On the other hand, EC8 specifies the transverse reinforcement within the critical length of the column (four-legged D8 @ 80, for medium ductility demand, or four-legged D8 @ 60 for high ductility demand) to continue within the joint. Thus, depending on the ductility demand, the code specified amount is two or three times as large as the amount required in accordance with the CFP method.

8.7 Structural Wall

The wall in Fig. 8.19 is considered to be subjected to a horizontal load P acting at a distance of 3,100 mm from the wall base. For the geometric characteristics, longitudinal reinforcement and material properties indicated in the figure, the CFP method is used determine the transverse reinforcement required for preventing non-flexural failure of the wall.

8.7.1 CFP Design

Flexural capacity. This is calculated as discussed for the case of the columns in Sects. 8.2 and 8.3 by taking into account the contribution of all available longitudinal reinforcement. Thus, flexural capacity $M_f \approx 1,616 \times 10^6$ Nmm and the corresponding load-carrying capacity $P_f = M_f/a_v = 1,616 \times 10^6/3,100 \approx 522 \times 10^3$ N, where a_v is the distance of the applied load from the wall base (shear span). The calculated values of the internal forces developing in concrete and steel when

Fig. 8.19 Geometric characteristics, longitudinal reinforcement details and material characteristics of structural wall

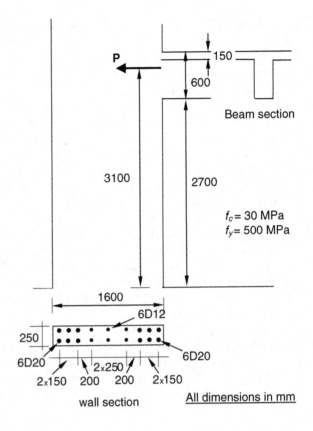

wall section   All dimensions in mm

the wall cross section attains its flexural capacity are shown in Fig. 8.20; the figure also shows the value and location of application of the resultant of the forces developing in the steel bars in tension.

Physical model. The wall is essentially a cantilever which is constructed as discussed in Sect. 6.2.1. Since the shear span-to-depth ratio $a_v/d = 3{,}100/979 \approx 3.17 > 2.5$, the wall is characterised by type II behaviour and, hence, its model is as illustrated in Fig. 8.21. As discussed in Sect. 2.3, for type II behaviour, brittle failure may either occur in the region of location 1 where the horizontal and inclined elements of 'frame' join (at a distance equal to $2.5d = 2.5 \times 979 \approx 2{,}447.5$ mm from the point of application of the applied load or at a distance equal to $a_v - 2.5d = 3{,}100 - 2{,}447.5 = 652.5$ mm from the wall base) or in the region of location 2 where the bending moment and shear force attain their largest values. [The causes of these types of behaviour are fully discussed in Chap. 2 (Sect. 2.3)].

Transverse reinforcement at locations 1. From expression 3.9, the 'shear' force that can be sustained without the need of transverse reinforcement in the region of location 1 is $V_{II,1} = 0.5f_tbd = 0.5 \times 2.37 \times 250 \times 979 = 290{,}029$ N < 5 22×10^3 N. Thus, there is a need for transverse reinforcement ($A_{sv,1}$) capable of sustaining the whole transverse tensile force $T_{II,1} = V_f = 522 \times 10^3$ N developing

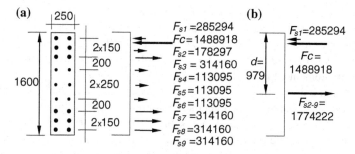

dimensions in mm and forces in N

Fig. 8.20 Internal forces developing in concrete and steel when the wall cross section attains its flexural capacity. **a** Tensile forces in each steel bar. **b** Tensile force resultant and location of application

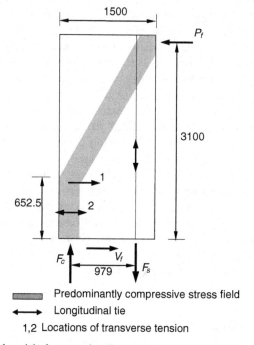

◼ Predominantly compressive stress field

⟷ Longitudinal tie

1,2 Locations of transverse tension

Fig. 8.21 Physical model of structural wall

within a length $2d = 2 \times 979 = 1{,}958$ mm, i.e. $A_{sv,1} = 522 \times 10^3/500 = 1{,}044$ m m^2. This amount of reinforcement is placed within a length extending to a distance $d = 979$ on either side of location 1, the latter, as discussed earlier, lying at a distance equal to 652.5 mm form the wall base. This reinforcement is provided in the form of eleven two-legged 8 mm diameter stirrups at spacing of 175 mm, with a total cross section of $1{,}105.84 > 1{,}044$ mm^2.

Transverse reinforcement at locations 2. From expression 3.13, the 'shear' force that can be sustained without the need of transverse reinforcement in the region of location 2 is $V_{II,2} = F_c^*[1 - 1/(1 + 5|f_t|/f_c)] = 1{,}488{,}918 \times [1 - 1/(1 + 5|2.37|/30)] = 421{,}593$ N $< V_f = 522 \times 10^3$ N. Thus, there is need for transverse reinforcement capable of sustaining the tensile stresses developing in the region of location 2 due to the loss of bond between concrete and the longitudinal reinforcement. The amount of reinforcement required for this purpose is calculated through the use of expressions 4.2 and 4.3. From expression 4.2, the nominal value of the transverse tensile stresses developing within the region of location 2 (extending from location 1 (i.e. the location of the joint of the vertical and inclined elements of the 'frame') to the wall base) is $\sigma_t = 30/[5 \times (1{,}488{,}918/522 \times 10^3 - 1)] \approx 3.23$ MPa, whereas, from expression 4.3a, the tensile stress resultant developing in the direction of the wall length within this region (with a length of 652.5 mm) is $T_{II,2v} = 3.23 \times 250 \times 652.5/2 \approx 263{,}447$ N. The amount of transverse reinforcement required to sustain $T_{II,2v}$ is $A_{sv,2v} = 263{,}447/500 = 526.9$ mm^2, i.e. six two-legged 8 mm diameter stirrups (with a total cross section of $603.18 > 526.9$ mm^2) placed at a 100 mm spacing. This amount of reinforcement is larger than the amount required within the same distance in order to safeguard against failure at location 1 and therefore the former replaces the latter within this region. On the other hand, from expression 4.3b, the tensile stress resultant developing across the direction of the wall length within the region of location 2 is $T_{II,2h} = 3.23 \times 142 \times 652.5/2 \approx 149{,}638$ N. The amount of transverse reinforcement required to sustain $T_{II,2h}$ is $A_{sv,2h} = 149{,}638/500 = 299.28$ mm^2, i.e. six 8 mm diameter bars (with a total cross section of 301.59 mm^2) which is covered by the legs of the stirrups $A_{sv,2v}$ across the wall length within the wall edges.

Total transverse reinforcement. Therefore, the transverse reinforcement required to safeguard against brittle non-flexural types of failure comprises two-legged 8 mm diameter stirrups placed throughout the wall height at spacing of (a) 100 mm within the region extending from the wall base to the cross section at a distance of 652.5 mm from the base, (b) 175 mm between the latter cross section and that at a distance of 1,631.5 and (c) 300 mm in the remainder of the wall, the latter being nominal reinforcement (see Fig. 8.22).

8.7.2 EC2/EC8 Design

As discussed in Sect. 7.3, the EC2/EC8 provisions for the design of earthquake-resistant RC structural walls specify a reinforcement arrangement comprising one part forming "concealed column (CC)" elements usually extending throughout the wall height along the its vertical edges and the other consisting of a set of grids of uniformly distributed vertical and horizontal bars, within the wall web, arranged in parallel to the wall large side faces. The CC elements are intended to impart to the walls the code specified ductility, whereas the wall web is designed against the occurrence of "shear" failure, before the wall flexural capacity is exhausted, the latter being assessed by allowing for the contribution of all vertical reinforcement which, for purposes of comparison, is considered to be that of the wall designed in accordance with the CFP method.

The specified ductility is considered to be achieved by confining concrete within the CC elements through the use of a dense stirrup arrangement—thus increasing both the strength and the strain capacity of the material. However, assuming moderate ductility demand and a curvature ductility ratio $\mu_\varphi = 4$, the use of expressions 5.20 in clause 5.4.3.4.2 of EC8 yields values of clear spacing equal to 10, 15 and 25 mm for the cases of stirrup diameters of 8, 10 and 12 mm, respectively, with all these solutions been inapplicable, since they result to severe reinforcement congestion.

On the other hand, shear failure is prevented by providing an amount of horizontal web reinforcement capable of delaying shear failure before the flexural capacity is exhausted; this reinforcement is found to be equal to two-legged 8 mm diameter stirrups at a 336 mm spacing (the latter rounded up to 300 mm for practical purposes) through the use of expressions 6.5–6.12 as described in clause 6.2.3 of EC2. The code specified reinforcement details are shown in Fig. 8.23.

And yet, comparing the code specified transverse reinforcement in Fig. 8.23 with that in Fig. 8.22 resulting from the application of the CFP method shows that, in accordance with the latter method, the code specified amount of transverse reinforcement within the CC elements is not only inapplicable, but also unnecessary, whereas the code specified reinforcement within the web of the wall is insufficient for safeguarding against a 'shear' type of failure.

Fig. 8.22 Reinforcement details of structural wall in Fig. 8.19 in accordance with CFP method

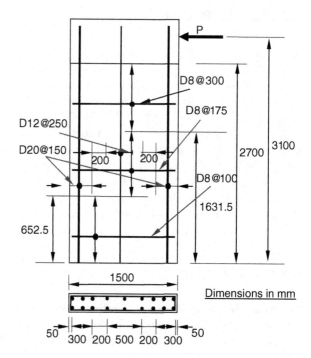

Fig. 8.23 Reinforcement
details of structural wall in
Fig. 8.19 in accordance with
EC2/EC8 provisions

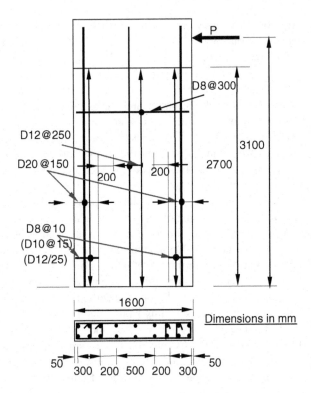

Fig. 8.23 Reinforcement
details of structural wall in
Fig. 8.19 in accordance with
EC2/EC8 provisions

8.8 Flat Slab Punching

A flat plate floor with the column layout indicated in Fig. 8.24 has a thickness
$h = 200$ mm and is supported by 500 mm square columns spaced 7,000 mm on
centres each way. The values of the top flexural reinforcement ratio in the column
and middle strips at the slab's cross section through the columns' axes of symmetry are $\rho_{t,cs} = 0.0081$ and $\rho_{t,ms} = 0.0026$, respectively, whereas those of the
bottom flexural reinforcement ratio in these strips at their cross section at mid
span between successive columns are $\rho_{b,cs} = 0.0032$ and $\rho_{b,ms} = 0.0021$, respectively. Check the adequacy of the slab resisting punching at a typical interior column before the slab's flexural capacity is reached, and provide reinforcement, if
needed. The average effective depth of the slab $d = 170$ mm, the compressive
strength of concrete $f_c = 20$ MPa and the yield stress of the steel $f_y = 500$ MPa.

8.8.1 CFP Design

Load-carrying capacity corresponding to flexural capacity. Consider the portion between successive columns of a longitudinal strip including a column strip
and half middle strip on either side of the column strip. The flexural capacities of

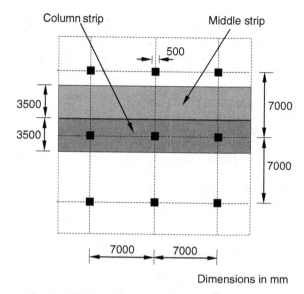

Fig. 8.24 Two-way flat plate floor

this strip's cross sections through a column and at mid span are calculated as indicated in Figs. 8.25 and 8.26.

Having calculated the values of flexural capacity at the above cross sections of the longitudinal strip considered, and assuming a uniformly distributed load, then, the maximum load that can be sustained per square metre by the slab before failing in flexure may be obtained by considering the bending moment diagram of the strip considered as indicated in Fig. 8.27. The figure also provides an indication of the location of the distance of the point of zero bending moment from the nearest support (column's axis).

Checking for punching. Punching may occur in the region of the slab enclosed by the geometric locus of the points of zero bending moment encompassing a column. For the case of the square column layout of the slab in Fig. 8.24,

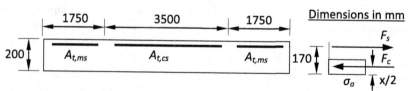

$A_{t,ms} = \rho_{t,ms} \times 1750 \times 170 = 0.0026 \times 1750 \times 170 = 773.5$ mm^2

$A_{t,cs} = \rho_{t,cs} \times 3500 \times 170 = 0.0081 \times 3500 \times 170 = 4819.5$ mm^2

For $f_c = 20$ MPa, $f_t = 1.58$ MPa (see expression 3.2(a)). Thus, $\sigma_a = 20 + 5 \times 1.58 = 27.9$ MPa

$F_s = (4819.5 + 2 \times 773.5) \times 500 = 3183250$ N; $F_c = 7000 \times 27.9x = 195300x$

From $F_s = F_c$, $x \approx 16$ mm and $z = 170 - 16/2 = 162$ mm. Thus, $M_f = F_s z = 3183250 \times 162 \approx 515.69$ kNm

Fig. 8.25 Calculation of flexural capacity of column strip at a cross section through a column

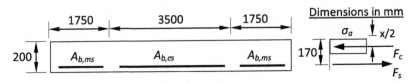

$A_{b,ms}=\rho_{b,ms}$ x1750x170=0.0021 x1750x170=624.75 mm²
$A_{b,cs}=\rho_{b,cs}$ x3500x170=0.0032x3500x170=1904 mm²
For f_c =20 MPa, f_t =1.58 MPa (see expression 3.2(a)). Thus, σ_a=20+5x1.58=27.9 MPa
F_s=(1904+2x624.75)x500=1576750 N; F_c=7000x27.9x= 195300x
From $F_s=F_c$, x≈8 mm and z=170-8/2=166 mm. Thus, $M_f=F_s z$=1576750x166≈261.74 kNm

Fig. 8.26 Calculation of flexural capacity of column strip at a cross section at mid span between successive columns

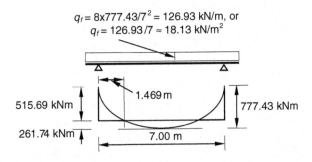

Fig. 8.27 Calculation of slab's load-carrying capacity corresponding to flexural capacity

this geometric locus forms a circle (see Fig. 5.2) with centre the axis of symmetry of an interior column and radius $r = 1,469$ mm (see Fig. 8.27).

The physical model of the above region is depicted in Fig. 8.28. As discussed in Sect. 5.2, punching may initiate at any of the following locations: (1) along the geometric locus of the points of zero bending moment where, as discussed in Sect. 6.2.1, the slab portion considered is linked to the remainder of the slab through the development of an internal support marked as '3' in the figure, (2) along the geometric locus of the location of change in the compressive force path, marked as '1' in the figure, with locations 1 being situated at a distance of $2.5d = 2.5 \times 170 = 425$ mm from the geometric locus of the points of zero bending moment, and (3) in the regions of strips along the axis of symmetry of the column's cross section, adjacent to the column-slab interfaces, marked as locations '2' in the figure.

The resultant $(T_{3,f})$ of the transverse tensile stresses developing along the perimeter of the slab's portion depicted in Fig. 8.28 is numerically equal the load (P_f) acting on this portion when flexural capacity is attained, i.e. $T_{3,f} = P_f = 18.1$ $3 \times (7^2 - \pi \times 1.469^2) = 765.46$ kN. On the other hand, the value that can be sustained $T_{u,3} = 0.5 \times 1.58 \times \pi \times [(1,469 + 170)^2 - (1,469 - 170)^2] = 2,479.18$ k $N > T_{3,f} = 765.46$ kN. Therefore, punching cannot initiate in this region.

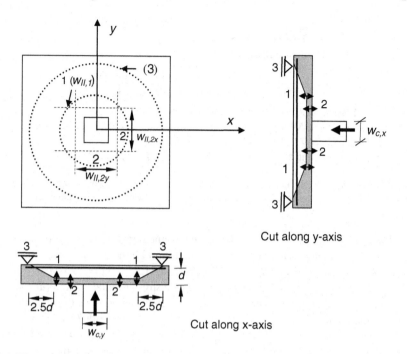

Fig. 8.28 Physical model of portion of the slab in Fig. 8.24 in the region of an interior column

The resultant ($T_{1,f}$) of the transverse tensile stresses developing along the geometric locus of locations '1' is numerically equal to the shear force resultant ($V_{1,f}$), i.e. $T_{1,f} = V_{1,f} = 18.13 \times [7^2 - \pi \times (1.469 - 0.425)^2] = 826.29$ kN, whereas the value that can be sustained is $T_{II,1} = 0.5 \times 1.58 \times \pi \times [(1,044 + 170)^2 - (1,044 - 170)^2] = 1,761.92$ kN $> T_1 = 826.29$ kN. Therefore, punching also cannot initiate in this region.

As discussed in Sect. 5.2.1, punching in the vicinity of the column is resisted by slab strips along the axes of symmetry of the column's cross section with width defined by expressions 5.1 and 5.2. From expression 5.2, $\lambda_1 = 2 - 100 \times 0.0081 \times 500/500 = 1.19$ and $\lambda_2 = 1$. Replacing λ_1 and λ_2 with their values, expression (5.1) yields the strip width $w_{II,2} = 500 + 2 \times 1.19 \times 170 = 904.6$ mm. The strip flexural capacity at its interface with the column and the corresponding internal forces F_c and F_s sustained by concrete and steel, respectively, are calculated as indicated in Fig. 8.29. Then, from expression 5.3, the punching force that can be sustained by the strips $P_{II,2} = 4V_{II,2} = 4 \times 622,817 \times [1 - 1/(1 + 5 \times 1.58/20)] = 705,413$ N. On the other hand, when the flexural capacity is attained, the punching force acting on the slab $P_f = 18.3 \times (7^2 - 0.5^2) = 892,125$ N $> P_{II,2} = 705,413$ N. Thus, reinforcement is required for preventing punching before flexural capacity is attained.

Reinforcement for safeguarding against punching. Each of the four strips should be reinforced so as to be capable of sustaining a shear force V_f equal to one quarter of the total acting punching force P_f, i.e. $V_f = P_f/4 = 892,125/4 = 223,031.2$

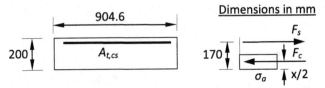

$A_{t,cs}=\rho_{t,cs}$ x904.6x170=0.0081x3500x170=1245.63 mm^2
For f_c =20 MPa, f_t =1.58 MPa (see expression 3.2(a)). Thus, σ_a=20+5x1.58=27.9 MPa
F_s=1245.63x500=622817 N; F_c=904.6x27.9x= 25238.34x
From $F_s=F_c$, x≈25 mm and z=170-25/2=157.5 mm. Thus, $M_f=F_s z$=622817x157.5≈98.1 kNm

Fig. 8.29 Calculation of flexural capacity (M_f) of column strip resisting punching at its intersection with column and corresponding internal forces sustained by concrete (F_c) and steel (F_s)

5 N. The transverse tensile stresses developing in the compressive zone of the strips when flexural capacity is attained are calculated through the use of expression (4.2). i.e. $\sigma_t = f_c/[5(F_c/V_f - 1)] = 20/[5 \times (622{,}817/223{,}031.25 - 1] = 2.23$ MPa.

Thus, from expression 4.3a, the vertical tensile stress resultant developing within the compressive zone of the strip between the cross sections at the column-slab interface and the location of change in the compressive force path $T_{II,2,v} = 0.5 \times 2.2$ $3 \times 904.6 \times [(1{,}469 - 2.5 \times 170) - 500/2] = 800{,}851.43$ N, whereas, from expression 4.3b, the horizontal tensile stress resultant developing within the compressive zone of the same portion of the strip $T_{II,2,h} = 0.5 \times 2.23 \times 25 \times [(1{,}469 - 2.5 - 1$ $70) - 500/2] = 22{,}132.75$ N. Ignoring the contribution of concrete (see Sect. 5.3), the total amount of reinforcement required to sustained these forces is $A_{sv,v} = 800{,}8$ $51.43/500 = 1{,}601.7$ mm^2 i.e. thirty two 8 mm diameter bars with 200 mm spacing across and 100 mm spacing along the strips and $A_{sv,h} = 22{,}132.75/500 = 44.27$ mm ^{2} i.e. one 8 mm diameter bar. The resulting reinforcement arrangement is, for clarity purposes, depicted in one of the four strips shown in Fig. 8.30.

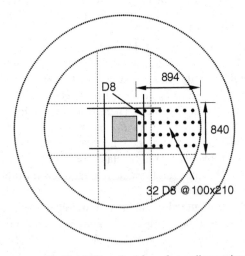

Fig. 8.30 Reinforcement specified by CFP method for safeguarding against punching

8.8.2 EC2 Design

The reinforcement specified by EC2 for safeguarding against punching is shown in Fig. 8.31a and b. As for the case of Fig. 8.30, for clarity purposes only the reinforcement placed within one of the four integral beams resisting punching is shown in Fig. 8.31b. Comparing the reinforcement details shown in Figs. 8.30 and 8.31 indicates that the total amount of reinforcement specified by the CFP method is 128 × D8 vertical + 4 × D8 horizontal bars = 6,635 mm^2 which is larger than both the 96 × D10 vertical bars = 4,825 mm^2 specified by EC2 within the integral beams and the 80 × D8 radially placed vertical bars = 4,021 mm^2 which is the alternative reinforcement arrangement also specified by EC2. It should be noted that, in contrast with EC2, CFP also specifies horizontal reinforcement in

Fig. 8.31 **a** Radial reinforcement specified by EC2 for safeguarding against punching. **b** Reinforcement specified within integral beams by EC2 for safeguarding against punching

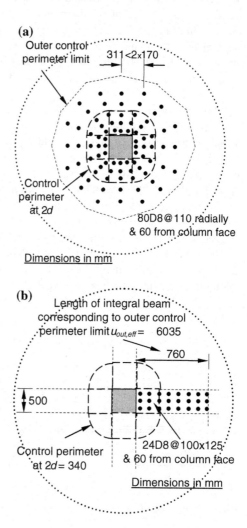

(a)

Outer control perimeter limit   311 <2×170

Control perimeter at 2d

80D8@110 radially & 60 from column face

Dimensions in mm

(b)

Length of integral beam corresponding to outer control perimeter limit $u_{out,eff}$ =   6035

760

500

Control perimeter at 2d = 340

24D8@100×125 & 60 from column face

Dimensions in mm

the compressive zone across the integral beam as indicated in Fig. 8.30. It should also be noted that, as discussed in Sect. 5.4.3, even if the code specified reinforcement were to increase to the amount of the reinforcement resulting from the use of the CFP method, the former would not be as effective as the latter, with this being indicative of the significant contribution of horizontal reinforcement specified by the CFP method to punching resistance.

8.9 Prestressed Concrete Beam

Figure 8.32 shows a prestressed concrete beam which is one of the specimens whose behaviour was investigated in a research programme [1] concerned with the verification of the validity of the initial version of the proposed method [1.1]. The beam, which is subjected to the loading arrangement also shown in the figure, is constructed with concrete with $f_c = 44$ MPa, tendons with a maximum sustained stress $f_{pu} = 1,900$ MPa and total cross-sectional area A_p 205.4 mm^2. Specify the amount of transverse reinforcement required for safeguarding against a brittle type of failure.

Flexural capacity. From Fig. 8.33, the equation $F_c = bx\sigma_c = F_s$ (where x is the depth of the compressive zone, $\sigma_c = f_c + 5f_t = 44 + 5 \times 3.29 = 60.45$ MPa (see expressions 3.2a and 3.3), and $F_s = A_p f_{pu} = 205.4 \times 1,900 = 390,260$ N) yields $x = 32.28$ mm. Hence, the lever arm of the internal longitudinal forces is $z = d - x/2 = 240 - 32.28/2 = 223.86$ mm, and thus the flexural capacity $M_f = 390,260 \times 223.86 = 87\ 363\ 604$ Nmm.

Shear force diagram. From the equilibrium conditions of any of the two halves of the beam, when it flexural capacity is attained, the values of each of the applied loads and the reaction are found to be $P_f = 16.05$ kN and $R_f = 3P_f = 3 \times 16.05 = 48.15$ kN, respectively. For these values of P_f and R_f, the shear-force diagram is as indicated in Fig. 8.34. From this diagram, the design shear force, within the portion of the beam between the support and the point load closest to it, is $V_f = 48.15$ kN.

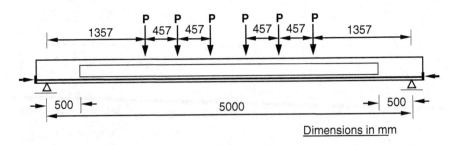

Fig. 8.32 Prestressed-concrete beam under a six-point loading arrangement

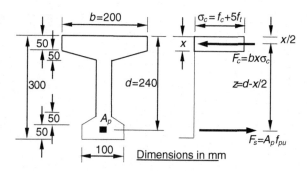

Fig. 8.33 Geometric characteristics and internal actions at mid cross section of prestressed-concrete beam in Fig. 8.32 when its flexural capacity is attained

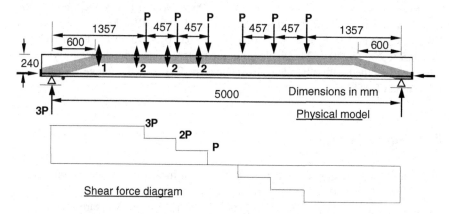

Fig. 8.34 Physical model of beam in Fig. 8.32 and shear force diagram

Physical model. Although the beam is subjected to six-point loading, the present case is treated as a case of two-point loading since, as indicated in Fig. 8.34, the six loads are applied in the middle portion of the beam, symmetrically about the mid-span cross-section, with the remainder of the beam comprising two large shear spans, each of length $a_v = 1{,}357$ mm. The location of the horizontal and inclined elements of the 'frame' is at a distance $2.5d = 2.5 \times 240 = 600$ mm from the closest support. As this location lies within the shear span, the beam should be characterised by either type I or type II behaviour depending on whether or not failure at locations 1 and 2 occurs under values of the shear force smaller than that corresponding to flexural capacity.

Checking for failure at locations 1 and 2. From expression 3.9, the shear force that can be sustained without the need of transverse reinforcement in the region of location 1 is $V_{II,1} = 0.5 \times 3.29 \times 200 \times 240 = 78{,}960$ N $> V_f = 48{,}150$ N. Also, from expression 3.13, the 'shear' force that can be sustained without the need of transverse reinforcement in the region of location 2 is $V_{II,2} = 390{,}260 \times [1 - 1/(1 + 5|3.29|/30)] = 138{,}208$ N $> V_f = 48{,}150$ N. Therefore, there is no need for transverse reinforcement, other than a nominal amount in the form of stirrups at a

Fig. 8.35 Nominal stirrup
arrangement

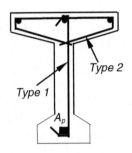

Type 2

Type 1

A_p

spacing of up to $0.5d$, capable of sustaining transverse stresses equal to 0.5 MPa. One stirrup with a 6 mm diameter at 120 mm cc spacing with $f_y = 460$ MPa is capable of sustaining $(\pi \times 6^2/4) \times 460 = 13,006$ N. As the transverse stress resultant of a uniformly distributed tensile stress of 0.5 MPa developing within a length of 240 mm is $T_w = 0.5 \times 50 \times 240 = 6,000$ N within the beam web and $T_f = 0.5 \times 200 \times 240 = 24,000$ N within the flange, a one-legged stirrup (type 1) in the web combined with a two-legged stirrup (type 2) within the flange are placed as indicated in Fig. 8.35.

8.10 Square Footing

A column 500 mm square, with $f_c = 30$ MPa, reinforced with eight 25 mm diameter bars of $f_y = 500$ MPa, supports a dead load of 1,000 kN and a live load of 800 kN. The allowable soil pressure q_a is 270 kN/m^2. Design a square footing with a base 1.5 m below grade, using $f_c = 30$ MPa and $f_y = 500$ MPa.

Preliminary design. Since the space between the bottom of the footing and the surface will be occupied partly by concrete and partly by soil (fill), an average unit weight of 20 kN/m^3 will be assumed. The pressure of this material at the 1.5 m depth is $1.5 \times 20 = 30$ kN/m^2, leaving a bearing pressure of $q_e = 270 - 30 = 240$ kN/m^2 available to carry the column service load. Hence, the required footing area $A_{req} = (1,000 + 800)/240 = 7.5$ m^2. A base 2.8 m square is selected, furnishing a footing of 7.94 m^2, which differs from the required area by about 4.5 %.

Following current design practice regarding the minimum reinforcement areas and maximum spacing (see clauses 9.2.1.1 and 9.3.1.1 of EC2), the flexural reinforcement selected comprises seven 20 mm diameter bars in either direction at a 400 mm spacing (the latter being the maximum allowed) (see Fig. 8.36). For a footing depth $d = 500$ mm, the total cross-sectional ratio of the above reinforcement in either direction is $\rho = 7 \times \pi \times (20^2/4)/(2,800 \times 500) = 0.00157 > \rho_{min} = 0.0013$.

Check for flexural capacity. From Fig. 8.36, the flexural capacity of the footing cross section containing column-footing intersection is $M_f \approx 545$ kNm and thus $q_f = 2 \times 544.502/(2.8 \times 1.15) \approx 339$ kN/m$^2 > q_e = 240$ kN/m^2.

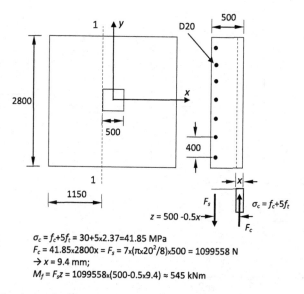

$$\sigma_c = f_c + 5f_t = 30 + 5 \times 2.37 = 41.85 \text{ MPa}$$
$$F_c = 41.85 \times 2800x = F_s = 7 \times (\pi \times 20^2/8) \times 500 = 1099558 \text{ N}$$
$$\rightarrow x = 9.4 \text{ mm};$$
$$M_f = F_s z = 1099558 \times (500 - 0.5 \times 9.4) \approx 545 \text{ kNm}$$

Fig. 8.36 Footing design details and calculation of flexural capacity

Modelling of footing. The footing of Fig. 8.36 may be viewed as an inverted flat slab subjected to a uniform pressure at its interface with the ground as a result of the patch load imposed upon it by the column. As for the case of a flat slab, punching is considered to be resisted by four strips fixed at the column's sides and extending along the axes of symmetry of the column's cross section; the pressure acting on the strips may be replaced by a point load (qa), acting at mid span, which is equivalent to the real load in that, not only is it equal to it and causes the same bending moment at the strip's fixed end, but also causes internal actions which, although more critical, do not exhibit a significant deviation from their real counterparts (see Fig. 8.37)

For the equivalent point load qa indicated in Fig. 8.37 the strip is modelled as indicated in Fig. 8.38. From Fig. 8.36, the shear span-to-depth ration $a_v/d = (1{,}150/2)/500 = 1.15$ and, hence, the strip is characterised by type III behaviour. From expression (5.2), $\lambda_1 = 2 - 100\rho f_y/500 = 2 - 100 \times 0.001$ $57 \times 500/500 = 1.843$ and $\lambda_2 = 1$; thus, from expression (5.1), $w_{II,2} = w_c + 2\lambda d = 500 + 2 \times 1.843 \times 500 = 2{,}343$ mm. The strip's flexural capacity is obtained from expression 3.14 in which $M_{III} = M_{II}^{(2.5d)} + (M_f - M_{II}^{(2.5d)})$ $(2.5d - a_v)/(1.5d)$, where $M_f \approx 390$ kNm is obtained as indicated in Fig. 8.38, $M_{II}^{(2.5d)} = (2.5d)\min(V_{II,1},V_{II,2})$, with $V_{II,1}$ and $V_{II,2}$ resulting from expressions 3.9 and 3.13, respectively, i.e., for $V_{II,1} = 0.5 \times 2.37 \times 2{,}343 \times 500 \approx 1{,}388$ kN and $V_{II,2} = 785{,}400 \times [1 - 1/(1 + 5 - 2.37/30)] = 222.39$ kN, $M_{II}^{(2.5d)} = 2.5 \times 0.5 \times 222.39 = 277.99$ kNm and, hence, $M_{III} = 277.99 + (390 - 277.99)$ $(2.5 - 1.15)/1.5 = 378.8$ kNm, the latter corresponding to $V_{III} = 378.8/$ $(1.15 \times 0.5) = 658.78$ kN. Thus, the four strips sustain a load $P_{III} = 4\,V_{III} = 4 \times$

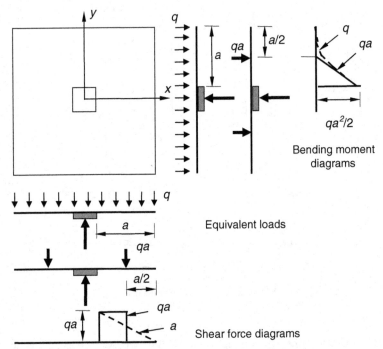

Fig. 8.37 Footing strips and equivalent loads and internal force diagrams

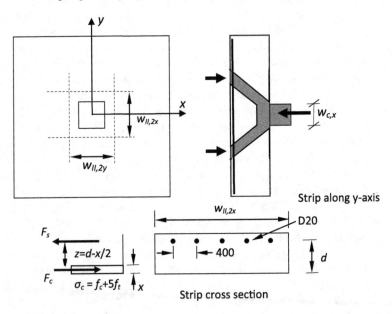

$\sigma_c = f_c + 5f_t = 30 + 5 \times 2.37 = 41.85$ MPa
$F_c = 41.85 \times 2343x = F_s = 5 \times (\pi \times 20^2/8) \times 500 = 785398$ N $\rightarrow x = 8$ mm;
$M_f = F_s z = 785398 \times (500 - 0.5 \times 8) \approx 390$ kNm

Fig. 8.38 Physical model of footing and calculation of flexural capacity of strips

$658.78 = 2,635.12$ kN $> 1,800$ kN and hence the footing is capable of sustaining the acting force without the need of transverse reinforcement.

EC2 check for punching. The check is made at the control perimeter at $2d$ which has a length $4w_c + 2\pi d = 4 \times 500 + 2 \times \pi \times 500 = 5,141.59$ mm. From expression 6.47 (clause 6.4.4 of EC2), the punching shear resistance $V_{Rd,c} = \{0.18 \times [1 + (200/500)^{1/2}] \times [100 \times 0.00157 \times (30 - 8)]^{1/3}\} \times 5,141.59 \times 500 = 1,141,884$ N $\approx 1,142$ kN $< 1,800$ kN. Hence, the footing is not capable of sustaining the acting force without the need of transverse reinforcement. In order to avoid placing transverse reinforcement, the footing height is increased to 980 mm; this increases $V_{Rd,c}$ to $1,811,685$ N $\approx 1,811$ kN $> 1,800$ kN and, hence, the footing becomes capable of sustaining the acting force.

Reference

1. Seraj SM, Kotsovos MD, Pavlovic MN (1993) Compressive-force path and behaviour of prestressed concrete beams. Mater Struct 26(156):74–89 (RILEM)

Index

Printed in the United States
By Bookmasters